Computers in Fisheries Research

Computers in Fisheries Research

Edited by

Bernard A. Megrey
National Marine Fisheries Service
Alaska Fisheries Science Center
Seattle, USA

and

Erlend Moksness
Institute of Marine Research
Flødevigen Marine Research Station
Norway

CHAPMAN & HALL

London · Glasgow · Weinheim · New York · Tokyo · Melbourne · Madras

Published by Chapman & Hall, 2–6 Boundary Row, London SE1 8HN, UK

Chapman & Hall, 2–6 Boundary Row, London SE1 8HN, UK

Blackie Academic & Professional, Wester Cleddens Road, Bishopbriggs, Glasgow G64 2NZ, UK

Chapman & Hall GmbH, Pappelallee 3, 69469 Weinheim, Germany

Chapman & Hall USA, 115 Fifth Avenue, New York, NY 10003, USA

Chapman & Hall Japan, ITP-Japan, Kyowa Building, 3F, 2-2-1 Hirakawacho, Chiyoda-ku, Tokyo 102, Japan

Chapman & Hall Australia, 102 Dodds Street, South Melbourne, Victoria 3205, Australia

Chapman & Hall India, R. Seshadri, 32 Second Main Road, CIT East, Madras 600 035, India

First edition 1996

© 1996 Chapman & Hall

Typeset in 10/12pt Photina by Acorn Bookwork, Salisbury, Wiltshire

Printed in Great Britain by St Edmundsbury Press, Bury St Edmunds, Suffolk

ISBN 0 412 59550 8

A catalogue record for this book is available from the British Library

Library of Congress Catalog Card Number: 95-71230

∞ Printed on permanent acid-free text paper, manufactured in accordance with ANSI/NISO Z39.48-1992 and ANSI/NISO Z39.48-1984 (Permanence of Paper).   [Paper = Magnum, 70 gsm]*

Contents

Contents

Preface

In 1989 we were asked by Dr Vidar Wespestad (National Oceanographic and Atmospheric Administration, National Marine Fisheries Service, Alaska Fisheries Science Center, Seattle, USA) to prepare and convene a session at the 1992 World Fisheries Congress in Athens, Greece on computer applications in fisheries. We agreed that the idea was a good one and the computer session turned out to be very successful. The computer session was organized in three parts: training classes, informal demonstrations accompanied by posters, and oral presentations of scientific papers. We were both amazed by the high level of interest and the high quality of contributions presented at the paper session. Returning from the World Fisheries Congress, we suggested to the International Council for the Exploration of the Sea (ICES) in Copenhagen to hold a theme session on the topic 'Computers in Fisheries Research' at their statutory meeting the following year in Dublin, Ireland. The proposal was very positively received by ICES and we began organizing this new meeting with Dr John Ramster of the Ministry of Agriculture, Fisheries and Food, Fisheries Laboratory, Lowestoft, England. Based on our experience with the World Fisheries Congress, we expected a maximum of 15 titles would be submitted to the ICES theme session. Accordingly, the ICES symposium was originally allocated one half-day time slot. The response we received from the call for papers, however, exceeded our most optimistic expectations. A total of 62 abstracts were submitted. Consequently, the symposium schedule was expanded to fill 3 days. During the oral presentations approximately 100 people participated, and even more attended the poster and demonstration session.

Nigel Balmforth from Chapman and Hall Ltd., who attended the World Fisheries Congress, asked us to consider preparing an outline for a book on the topic of computer use in fisheries research. Based on our two recent experiences, we knew that the interest level in the international fisheries community was high. We saw first-hand that individuals are quick to realize the potential of computers in fisheries and more scientists were taking advantage of these new tools. These observations and experiences convinced us that there was a need for such a book and that the idea was timely. Some investigating indicated that a book that reviewed current and future computer trends in fisheries science applications did not exist so we decided to proceed. The chapters that follow are the fruition of this idea. The book is intended to provide a sampling of contemporary fisheries computer applications. Hopefully by examining the chapters that follow, scientists will have an opportunity to evaluate the suitability of different computer technology applications to their particular research situation thereby taking advantage of the experience of others.

Originally, Dr Philip Sluczanowski, of the South Australian Research and Development Institute, was asked and accepted the task of writing the final chapter. Philip was ideally suited to this task as he was quite a visionary in terms of how fisheries computing needed to develop. He was a strong advocate of using highly visual computer models to communicate fishery management issues to nonfishery individuals. However, he suddenly passed away in June 1994. May he rest in peace.

We believe this book will be of interest to a wide audience. It should be useful as background reading for graduate and undergraduate students taking courses in quantitative resource management, fisheries, or the application of computers to specific fisheries subdisciplines. Academic institutions with agriculture, fisheries or resource management programs, national fisheries laboratories, and library systems should find the book useful. The book should also prove useful to administrators, managers, research scientists, field biologists, university researchers, university teachers, graduate and undergraduate students, consultants, government researchers, and laypersons involved in the fisheries or natural resource disciplines.

We would like to thank all who participated in the ICES symposium and the World Fisheries Congress. We are very grateful for the enthusiastic response we received from all the chapter authors during the preparation of this book. The quality of the presentations that follow reflect the high level of skill and enthusiasm they bring to their work. We would finally like to acknowledge the numerous reviewers who unselfishly contributed their time and expertise to improve the expositions contained herein.

Bernard A. Megrey
Erlend Moksness
December 1994

Contributors

Kenneth G. Foote
Institute of Marine Research
Department of Marine Environment
PO Box 1870 Nordnes
N-5024 Bergen
Norway

Ray Hilborn
University of Washington
School of Fisheries WH-10
Fisheries Research Institute
Seattle, WA 98195
USA

Ferren MacIntyre
ETI, Institute for Taxonomic Zoology
University of Amsterdam
PO Box 4766
NL-1090 GT Amsterdam
The Netherlands

Geoff J. Meaden
33 St. Stephen's Road
Canterbury
Kent CT2 7JD
UK

Bernard A. Megrey
National Marine Fisheries Service
Alaska Fisheries Science Center
7600 Sand Point Way N.E.
BIN C 15700
Seattle, WA 98115
USA

Erlend Moksness
Institute of Marine Research
Flødevigen Marine Research Station
N-4817 His
Norway

Thomas T. Noji
Institute of Marine Research
Department of Marine Environment
PO Box 1870 Nordnes
N-5024 Bergen
Norway

Pierre Petitgas
ORSTOM
H.E.A. Laboratory
911 Avenue Agropolis
BP 5045
Montpellier 34032 cedex
France

Kenneth A. Rose
Oak Ridge National Laboratory
Environmental Sciences Division
PO Box 2008
Oak Ridge, TN 37831-6036
USA

Edward S. Rutherford
Institute for Fishery Research
212 Museums Annex Building
University of Michigan
Ann Arbor, MI 48109-1084
USA

Saul B. Saila
Graduate School of Oceanography
University of Rhode Island
Narragansett, RI 02882-1197
USA

Dennis SinghDermot
Oak Ridge National Laboratory
Environmental Sciences Division
PO Box 2008
Oak Ridge, TN 37831-6036
USA

Jeffrey A. Tyler
Oak Ridge National Laboratory
Environmental Sciences Division
PO Box 2008
Oak Ridge, TN 37831-6036
USA

Carl J. Walters
Fisheries Centre
University of British Columbia
2204 Main Hall
Vancouver
British Columbia V6T 1W5
Canada

Trademark notices

Chapter one

Introduction

Bernard A. Megrey
and Erlend Moksness

*It is unworthy of excellent men to lose hours like slaves in the labor of
calculation which could be relegated to anyone else if machines were used.*
Gottfried Wilhelm von Leibniz
(1646–1716)

*The operations of analysis are now capable of being executed by machinery
... As soon as an analytical engine exists, it will necessarily guide the
future course of science.*

Charles Babbage
(1792–1871)

The nature of scientific computing has changed dramatically over the past
couple of decades. The appearance of the personal computer in the early
1980s changed forever the landscape of computing. The two quotes above,
one by the father of calculus and the other by the father of modern digital
computers, are particularly relevant as they relate to the use of computers
in fisheries research.

Fisheries science as a discipline was slow to adopt personal computers on
a wide-scale basis, being well behind use in the business world. In the very
early stages this was mainly a budgetary constraint (Walters, 1989).
Machines were scarce and it was common for more than one user to share
a machine, which was usually placed in a public area. Nowadays, in many
modern fisheries laboratories, a personal computer in every office is the rule
rather than the exception. Use of computers and computer applications to
support fisheries and resource management activities is rapidly expanding,
as are the diversity of research areas where they are being applied. The
important role they play in contemporary fisheries research is unequivocal.
These trends, which are taking place throughout the world-wide fisheries
community, have produced significant gains in work productivity,

Computers in Fisheries Research.
Edited by Bernard A. Megrey and Erlend Moksness.
Published in 1996 by Chapman & Hall, London. ISBN 0 412 59550 8

increased our basic understanding of natural systems, helped fisheries professionals focus on working hypotheses, provided badly needed tools to
rationally manage scarce natural resources, and in general have encouraged clearer thinking and more thoughtful analysis of fisheries problems. If
Leibniz were alive today, he undoubtedly would be using a personal
desktop computer to examine concepts in mathematics and numerical
methods that extend far beyond the calculus he helped to shape.

The objective of this book is to provide a vehicle for fisheries professionals
to keep abreast of recent and future developments in the application of
computers in their specific area of research and to familiarize them with
new application areas. We hope to achieve the objective by identifying
experts from around the world to present overview papers on topic areas
that represent current and future trends in the application of computer
technology to fisheries research. The aim is to provide critical reviews on
the latest, most significant developments in selected topic areas that are at
the cutting edge of the application of computers in fisheries and resource
management.

Many of the topics in this book cover areas that were predicted in 1989
to be important in the future: image processing, stock assessment, simulation and games, and networking (Walters, 1989). The chapters that follow
cover these areas as well as others and suggest several emerging trends and
future directions that impact the role computers are likely to place in fisheries research.

1.1 FASTER COMPUTING PLATFORMS

We marvel at how quickly computer technology has advanced. The current
typical desktop computer, compared to the original monochrome 8 KB
random access memory (RAM), 4 MHz 8088 microcomputer or the original
Apple II, has improved several orders of magnitude in several areas. The
most notable of these are processing capability, color graphics resolution
and display technology, hard disk storage, and the amount of RAM. The
remarkable thing is that, since 1982, the cost of a high-end Intel-based
microcomputer system has remained in the neighborhood of US$3000.

Computer technology is changing at an ever rapid pace. Higher performance central processing units (CPU) require more logic circuitry and this
is reflected in steadily rising transistor densities. Transistor density, a rough
measure of power, doubles approximately every year and a half (*PC
Magazine*, 1994a). In 1982, the 8086 CPU had less than 29 000 transistors. In 1995, the P6 has 5.5 million (Figure 1.1). Also, the processing and
data handling capabilities continue to improve. At this writing the succes-
sor to the Pentium CPU, the sixth generation 133 MHz P6, is expected to be

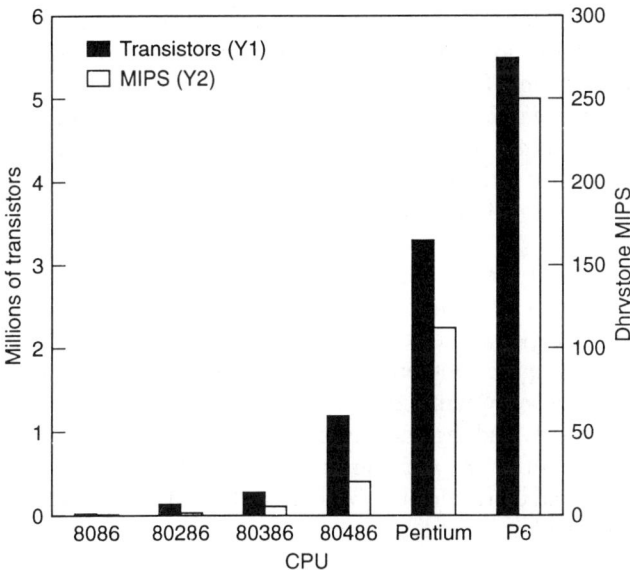

Fig. 1.1 Performance and design history of *x*86 microprocessors (*BYTE*, 1995a).

twice as powerful as the 100 MHz Pentium. The number of millions of instructions per second (MIPS) that Intel's top-of-the-line complex instruction set computer (CISC) CPU executed in 1983 was 1.2. In 1993 it was 112.0 and in 1995 it was 250 (Figure 1.1). The remarkable thing is that while the number of transistors per CPU has increased more than 100 times over the past 16 years, performance has increased more than 500 times (*BYTE*, 1995b). Newer reduced instruction set computer (RISC) chips, such as Digital's Alpha 21164 processor and the PowerPC, run 2–4 times faster than a top performance Pentium-based PC. Other CPUs that are at the heart of powerful UNIX-based workstations such as the Sun Ultra-SPARC, Mips R10000, and the Hewlett-Packard P-8000 are comparable. Analysts that use large databases, scientific visualization applications, statistics, and simulation modeling need as many MIPS as they can get. The more powerful computing platforms described above will enable us to perform analyses that we could not perform earlier (Chapters 7 and 8).

Three years from now CPUs will be four times faster than they are today and multiprocessor designs should be commonplace. Standard RAM configurations will continue to increase as well. In 3–5 years 64–256 Mb of dynamic RAM will be available and machines with 64 Mb of RAM will be typical. Permanent hard disk storage is also increasing in capacity and speed. Fast multigigabyte hard disks are commonplace and they are continuing to get smaller and faster. Newer storage media such as read–write capable CD-ROM and write once read many times (WORM) drives will eventually displace floppy disks as the storage medium of choice. In addition, improvements in semiconductor performance will enable dramatic advances in graphics technology. In 3–5 years, support for 24-bit color, full 3-D acceleration, broadcast quality video, and full-motion near-lifelike virtual-reality capabilities will be commonplace as well (*PC Magazine*, 1994b).

Another recent trend that has emerged is the appearance of powerful portable computer systems. The first portable computer systems were large and heavy, and portability often came at a cost of reduced performance. Today current laptop, notebook, and subnotebook designs are often comparable to desktop systems in terms of their processing power, hard disk and RAM storage, and graphic display capabilities. It is not unusual, when attending a scientific or working group meeting, to see most participants arrive with their own portable computers loaded with data and scientific software applications.

1.2 POWERFUL SOFTWARE

Spreadsheet applications were probably the first microcomputer software to be embraced by fisheries scientists in a significant way. The most significant impact of the spreadsheet was that it moved use of computers from the hands of the programmer to the regular biologist. No longer were long hours of tedious programming required to get data into and out of the computational heart of a computer program. Calculations were interactive and instantaneous. And on top of that, spreadsheets provided easy-to-use interactive graphics that previously were only available by submitting files of long cryptic commands to off-line graphics devices such as drum plotters. We remember attending many fisheries meetings and coming to recognize the familiar graphic output produced by Lotus 1-2-3.

Software functionality and growing feature sets have advanced in lockstep with improvements in computer hardware performance and expanded hardware capability. Using the computer is much easier now that graphically oriented user interfaces such as System 7, MS Windows, and Xwindows are commonplace. No longer is the user forced to interact with the computer using obscure complicated commands.

Today's application software packages are extremely powerful. Scientific data visualization tools and sophisticated multidimensional graphing applications facilitate exploratory analysis of large and complex data sets and allow scientists to investigate patterns in their data that were difficult to examine several years ago. This trend enables users to focus their attention on interpretation and hypothesis testing rather than on the mechanics of the analysis. Software that permits the analysis of the spatial characteristics of fisheries data are becoming more common. Programs to implement geostatistical algorithms (Chapter 5) and geographic information system (GIS) software (Chapter 3) offer fisheries biologists the ability to consider this most important aspect of natural populations. Image analysis software (Chapter 6) also offers promise in the areas of pattern recognition. Even the lowly spreadsheet has evolved into a powerful analysis tool that, in one software application, offers the user database management, publication quality graphics, statistical testing procedures, nonlinear minimization algorithms, and much much more.

Highly specialized software such as neural networks and expert systems (Chapter 2) have received limited application to fisheries problems. But still, these tools offer exciting new areas of research previously unavailable to fisheries scientists. The marriage of powerful computer systems to remote sensing apparatus and other electronic instrumentation continues to be an area of active research and development (Chapter 2). The area of population dynamics, fisheries management, and statistical methodology software, long a mainstay of computer use in fisheries, continues to receive very much attention (Richards and Megrey, 1994; also see Chapter 7) and many types of analyses have been successfully converted from mainframe to microcomputer platforms (Edwards and Megrey, 1989).

1.3 BETTER CONNECTIVITY

Desktop computers can be expected to interconnect to a wider sphere of networks and exchange data easier. One resource that will facilitate this interconnectivity and ease of use is the Internet, a vast international network of computers that spans the entire world. Currently, the Internet connects 25 million users and is made up of 45000 networks in 137 countries using 3.8 million host computers (*PC Week*, 1994).

With Internet navigation tools such Mosaic, an Internet browser created by the National Center for Supercomputing Applications, users have simple-to-use point-and-click access to the World-Wide Web. This is an Internet information retrieval system with more than 2300 graphical multimedia databases of 'hyperlinked documents'. The hypertext links can easily take you around the world in a matter of minutes, providing users instant access to data and other information resources. The beauty of Mosaic is that it lets you get there just by clicking your mouse, so you need not be bothered with where the information is located.

Use of the Internet is growing rapidly. The current user base is doubling every year (*Time*, 1994) and the approximately 137 countries that are reachable by electronic mail is growing rapidly as well. The annual rate of growth for World Wide Web traffic is 341 634% (*SUNEXPERT Magazine*, 1994).

1.4 SUMMARY

Further miniaturization of the microcomputer, increased capacity both in terms of processing speed, memory and storage capacity, and at relatively stable prices, will be seen. Although these changes are eagerly anticipated, we probably will never be satisfied. As performance levels increase so too will the complexity of our work and our expectations of 'minimum system configuration' and 'adequate performance'. The desire for faster machines, better graphics, and larger hard disks will not go away in the near future.

Computer software will continue to provide expanded analysis capabilities to fisheries professionals. We now have microcomputer operating systems that surpass what was commonly available on large mainframe computers several years ago, communication tools that make world-wide communication not only possible but fun, great suites of consistent, powerful productivity applications, personal database applications with sophisticated development and data management tools, graphic programs that let you do things you didn't think were possible a couple of years ago, and wider availability of highly specialized fisheries analysis programs.

The Internet, other network, and wide-area network connectivity resources hold great promise to deliver global access to a vast and interactive knowledge base. In addition, the Internet will provide to the user a transparent connection to networks of information and, more importantly, people. This later capability will allow fisheries scientists to communicate with colleagues within the scientific community who they might not otherwise have the opportunity to meet as well as external clients (Schnute and Richards, 1994). Moreover, it will provide expanded capabilities to share databases, computer programs, and experience. Walters (1989) predicted such an expansion of networked microcomputers. Storage and sharing of fisheries database and historic information products is encouraging the systematic analysis of past fisheries information (Chapter 7) and facilitating interregional comparison of fisheries and experience. Common patterns and trends may emerge as these datasets are shared, made available for quick retrieval and analysis, and accompanied by companion datasets such as vast oceanographic and meteorological data libraries. Increased microcomputer connectivity holds great promise for research collaboration. In fact much of the communication for organizing this book was accomplished over the Internet. With its working prototype, already traveled by 20 million users around the world, the Internet is opening new frontiers never before imagined.

Computers, fisheries research, and the resulting synergism is indeed exciting. The following collection of papers, while not intended to be comprehensive, does characterize the breadth and sophistication of the application of computers to modern fisheries analysis. We believe that the topics covered here are a prelude to new future opportunities and we anticipate with enthusiasm the challenges ahead.

REFERENCES

BYTE (1995a) **20**(4), 48.

BYTE (1995b) **19**(6), 88.

Edwards, E.F. and Megrey, B.A. (1989) Rational and organization of this volume, in *Mathematical Analysis of Fish Stock Dynamics*, (eds E.F. Edwards and B.A. Megrey), American Fisheries Society Symposium 6, 1–2.

PC Magazine (1994a) **13**(21), 105.

PC Magazine (1994b) **13**(22), 153.

PC Week, (1994) Dec 26, 18.

Richards, L.J. and Megrey, B.A. (1994) Recent developments in the quantitative analysis of fisheries data. *Canadian Journal of Fisheries and Aquatic Science* **51**(12), 2640–1.

Schnute, J.T. and Richards, L.J. (1994) Stock assessment for the 21st century. *Fisheries* **19**(11), 10–16.

SUNEXPERT Magazine, (1994) August, 12.

Time (1994) July 25.

Walters, C.J. (1989) Development of microcomputer use in fisheries research and management, in *Mathematical Analysis of Fish Stock Dynamics*, (eds E.F. Edwards and B.A. Megrey), American Fisheries Symposium **6**, 3–7.

Chapter two

Guide to some computerized artificial intelligence methods

Saul B. Saila

2.1 WHAT IS ARTIFICIAL INTELLIGENCE?

Artificial intelligence is not a new concept, because since the late 1940s exciting developments have been taking place in making computers act in a way that one would consider intelligent. Artificial intelligence (AI) has been defined as a subfield of computer science which deals with machines and software that appear to think and solve problems. The overall goal of much of the work in artificial intelligence is to develop systems that perform like human beings. Whether or not these systems are really 'intelligent' is a philosophical question. For our purposes, the fact that these systems mimic human intelligence to some extent at least is sufficient. AI programs are programs that exhibit behavior normally identified with human intelligence, and until recently, not with computers. In particular, AI programs seem to grasp ideas and concepts. For example, some can understand natural languages, some can see or perform difficult assembly tasks, and others can infer new information from facts provided to the system. However, software that makes computers seem intelligent is not that different from any other kind of software.

Since AI is still a very rapidly evolving discipline, it is difficult at present to break it down into well defined basic categories. Rauch-Hindin (1988) has partitioned AI programs into three basic categories: expert (knowledge-based) systems and the tools to build them, natural language (everyday native language) systems, and perception systems for vision, speech, and touch. In the above context, expert systems are programs that contain the knowledge of human experts, encoded so that a computer can understand

Computers in Fisheries Research.
Edited by Bernard A. Megrey and Erlend Moksness.
Published in 1996 by Chapman & Hall, London. ISBN 0 412 59550 8

it and utilize it through a human-like encoding mechanism (inference engine) to solve problems in specific knowledge domains. Natural language systems include programs that understand the native language of the user, such as English. Some of these natural language systems act as interfaces to databases to allow users to query databases in fairly unconstrained English instead of a formal programming language. Perception systems such as computer vision systems can, for example, interpret visual scenes and decide if an object meets defined quality control standards, or move a robot to the proper location in order to perform some assembly task.

For the purposes at hand, the first category above (expert systems) is included but not the latter two categories. Instead, this guide includes neural networks, genetic algorithms, and simulated annealing as currently interesting and potentially valuable AI tools for fishery scientists. The rationale for this choice is briefly as follows. Since the inception of the term AI, its role has been broadly defined as developing a means for increasing our understanding of human cognition and for building machines that simulate human decision making in uncertain and imprecise environments. Much of the early work in AI was based on symbol manipulation and first-order logic. This resulted in some significant accomplishments with expert systems and to a lesser extent in natural language processing. However, AI has recently contributed more significantly to robotics, machine translation, and pattern recognition problems. These latter problems have involved numerical computations which include uncertainty and imprecision. Kosko (1992) has provided a lucid dynamic systems approach to machine intelligence which utilizes neural networks and fuzzy systems theory to address the above-mentioned classes of problems. It is believed that these will have considerable relevance to some future developments in fishery science.

2.2 KNOWLEDGE-BASED SYSTEMS (EXPERT SYSTEMS)

2.2.1 Background

The emergence of artificial intelligence technologies in fishery science comes at a time of precipitous changes in the concepts and operational tools of fishery science and management. Many of the major fish stocks of the world are overexploited and many of the traditional models and assessment techniques are being challenged. After years of continuing global yield increases and open access fisheries, fishers are facing decreasing yields and real challenges to open access policy in the major capture fisheries. Both fishers and fishery scientists in the next decade must become more expert managers of all aspects of fisheries. It is my belief that the numerical modeling methods often used for current decision making may fail because much of our understanding about the entire fishery system is qualitative and/or highly uncertain, which does not lend itself well to conventional mathematical representation.

Knowledge-based systems (expert systems) offer a possible alternative to mathematical models as a way to represent our knowledge about fisheries and to apply that knowledge to solve problems. They are not expected to become a panacea. However, it is believed that they can be a useful alternative for consideration because their use is an admission that our knowledge is incomplete and uncertain. It is believed that building and testing expert systems for some fishery problems is a move toward practicality. That is, in building and testing such systems we recognize that making decisions based on qualitative and sometimes incomplete understanding is still better than making decisions without any understanding. Fortunately, it is not necessary to make a choice because expert systems can include conventional numerical methods. However, the major focus of expert systems is based on reasoning and how the numerical tools can provide the information needed to make rational decisions.

2.2.2 What are expert systems?

Simply stated, expert systems are computer programs that rely on a body of knowledge to perform a relatively difficult task usually performed only by a human expert. Artificial intelligence (AI) achieved considerable success in the development of expert systems during the period from the mid-1960s to the early 1990s. For example, Waterman (1986) provided a guide which described the application of expert system technology to diverse disciplines, including accounting, agriculture, chemistry, electronics, environmental management, geology, law, mathematics, medicine, meteorology, physics, process control, and others. Warwick *et al.* (1993) divided problems suitable for solution by an expert system into two major groups: analysis and synthesis. Analysis applications (including diagnosis, prescription, or solution finding, selection, and advisory) involve deriving a small number of outcomes or decisions given a large number of factors or conditions. Expert systems solve analysis problems by calculations and heuristic rules to relate the required outcomes or decisions to given conditions. Synthesis problems are the reverse of analysis problems. These involve the derivation of the conditions or parameters necessary to achieve a given outcome as closely as possible, taking into account any constraints or requirements. Scheduling, design, and configuration problems are typical synthesis problems. Warwick *et al.* (1993) found that all eight of the problem types considered suitable for expert system applications (namely interpolation, prescription, prediction, design, planning, monitoring and control, diagnosis, and instruction) had been used in environmental management.

There seems to have been a rapid and continuing increase in the use of expert systems in this discipline. To provide some indication of the use of expert systems for environmental subjects, Davis and Clark (1989) list a bibliography of 209 references related to natural resources. Warwick *et al.* (1993) cited 98 references (by my count) related to expert systems in environmental management. Clearly, environmentalists have attempted to capitalize on the use of AI through expert system development. This does not seem to be the case with aquatic scientists dealing with fishery resources. Most of the limited references to expert systems in aquatic sciences relate to ocean engineering and physical oceanography. Specifically, these often involve control and maneuverability systems for surface and subsurface vehicles and prediction or classification of physical resources.

A search of *Aquatic Sciences and Fisheries Abstracts* from 1988 to September 1993 produced 39 references which involved some aspect of fisheries and expert systems. Table 2.1 lists those references which were judged to be most relevant for this chapter. It is evident from this table that the amount of research and development done with expert systems in fisheries is very limited in contrast to, for example, environmental management.

2.2.3 Expert system development

Many of the expert systems developed during the 1970s and 1980s were based on variants of the LISP programming language using mainframe or

Table 2.1 Some examples of expert systems used in fisheries

Reference	Summary
Aoki *et al.* (1989)	expert prediction system for anchovy
Bender *et al.* (1992)	fishery design
Bossu *et al* (1989)	sanitary control in fish culture expert system
Calisal *et al.* (1992)	fishing vessel design system
Dillon and Lutz (1991)	aquatic animal bioaccumulation
Fréon *et al.* (1994)	surplus production model
Froese *et al.* (1989)	expert system for fish identification
Froese and Schoefer (1987)	larval fish identification
Fuchs (1991)	fish and environment-related expert system
Hanfman *et al.* (1989)	REGIS: aquaculture information system
Lee *et al.* (1991)	risk and bioaccumulation system
Markert and Klausmeyer (1990)	sampling expert system
Rallet *et al.* (1990)	aquaculture and trout farm expert system
Recknagel *et al.* (1991)	water quality expert system
Ryan and Smith (1985)	fisheries management
Stagg (1990)	expert support system for fishery management
Stokoe and Gray (1990)	AQUASITE: coastal aquaculture site assessment
Stretta *et al.* (1990)	expert system for tuna fishery

minicomputers. Lehner and Barth (1985) were among the first to focus attention on microcomputer-based expert systems. This section relates almost exclusively to these kinds of systems, which were developed more recently.

The elements of a knowledge-based expert system are shown in Figure 2.1. The knowledge base may be perceived as a computerized reference or manual containing topic-specific facts, rules relating to these facts, and rules-of-thumb (heuristics) used by subject matter experts to solve problems in a specific subject area (domain). Rules are mostly composed in consultation with one or more experts, and prototype system performance is often evaluated on hypothetical cases applied to the system. The knowledge base, consisting of a list of potential conclusions, the features and the rule set, is usually taken to be complete when the subject matter expert is satisfied. It is important to remember that this form of knowledge is not a random collection of facts but a collection of facts related by some well defined overall structure.

The inference engine is a term used in artificial intelligence to describe an isolated part of a computer program that performs some defined function. Knowledge bases of all rule-based expert systems share a fundamental component – the production rule. In the simplest case a production rule has the form:

IF	< assertion 1 >
AND	< assertion 2 >
AND	...
AND	< assertion m >
THEN	< assertion $m + 1$ >
AND	...
AND	< assertion $m + n$ >

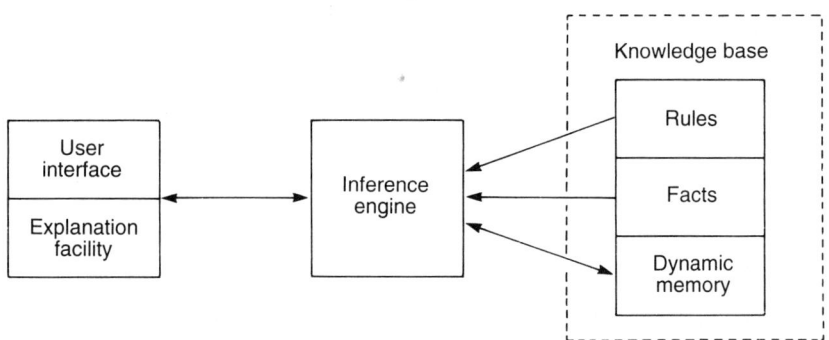

Fig. 2.1 A simple model of the structure of a knowledge-based expert system.

The *m* assertions in the IF part of the rule are called antecedents, and the *n* assertions in the THEN part of the rules are called consequents. Rules cannot contain direct references to other rules. Determining which rule to invoke as a logical consequence of another rule involves the inference engine. The form of the rule given above, in which the antecedents and consequents are each single statement assertions or propositions, implies that the expert system is based on propositional logic. Dynamic memory contains all the information that the expert system has gathered in pursuit of the solution. It contains information about rules and assertions. Information which must be stored includes the rules and whether they have already been used. Rules are tested when the truth values (i.e. true or false) of at least one of the antecedents is unknown. If all antecedents are known to be true, then the rule is invoked (instantiated), and the validity of the consequents is established. If one or more of the antecedents is known to be false, then the rule cannot be instantiated. When the truth values of the rule are established, the rule is marked as used.

The inference engine is the program code that contains the control logic for manipulating the rule base to make a conclusion or conclusions, that is, the part of the code that performs inference. In rule-based systems, the process of inference is often called chaining since rules are chained together in the inference process. Two types of chaining are forward, or data-driven chaining, and backward, or goal-driven chaining. A forward chaining inference engine starts with all the facts and repeatedly works through the rule-base adding new conclusions until no more can be added. Conclusions are reached by instantiating rules if and when all of their antecedents (the IF parts) have been verified. Backward, or goal-directed chaining, starts with a set of hypotheses or goals and attempts to piece together chains of rules to reach one or more members of the set. Backward chaining is intuitively more attractive because human reasoning (the formulation and testing of hypotheses) seems to be similar.

One of the features that separates an expert system from a traditional computer program is the ability of the expert system to work with symbolic information – information that does not involve numeric values. A standard program can easily accommodate quantitative rules-of-thumb, such as 'if the chi-square statistic is less than 3.84 with one degree of freedom, do not consider the results significant at the alpha 0.05 level.' However, it is harder to incorporate descriptive rules such as 'if any birds are sighted excitedly whirling in a small area near the surface of the sea, it usually means that a school of fish may be found in the vicinity.' Although expert systems are not the only form of computer programs which can work with symbolic information, declarative languages such as PROLOG and LISP provide an easy way to handle this problem.

2.2.4 Characteristics, strengths, and weaknesses of expert systems

In order to be effective, an expert system should have several characteristics. Four important characteristics are as follows:

(a) In order to add more knowledge and to improve performance, expert systems must have the capability to update their knowledge base regularly.
(b) Expert systems should have flexible problem-solving strategies in order to cope with many real-world problems.
(c) They should be able to solve most assigned problems correctly.
(d) They should be able to explain what they have done and why to the user.

Not all fields of knowledge are currently suitable for expert system applications. In general, expert system applications seem suitable for tasks which people perform better after considerable experience, and which are well defined and not too broad. A potential weakness of expert systems is that experts in a particular domain (discipline) may not be able to systematically communicate their knowledge so that it can be effectively incorporated into the knowledge base. Clearly, inadequate knowledge in any form can adversely affect the functioning of the system. In addition, poor engineering and testing, which creates an expert system which does not clearly match the performance of human experts, is also a potential weakness.

Many of the initially perceived barriers to the development of expert systems have been reduced substantially in recent years. The cost and size of computers large and powerful enough for expert system development is no longer a problem, due to the advent of fast microcomputers. Development time and the difficulty of learning new declarative languages have been substantially reduced by the wide availability of expert system shells, which greatly simplify the development of expert systems for specific discipline-related applications. The shell consists of those parts of the expert system that are independent of the expert's knowledge, usually the inference engine and the dynamic memory. Examples of expert systems developed from an expert system shell (Exsys^{TM}) include Dillon and Lutz (1991) and Lee *et al.* (1991).

The strengths of expert systems are believed to include the following:

(a) Expert systems provide a framework within which it is possible to apply algorithmic knowledge and to disseminate it quickly and effectively at low cost. Fishery systems are generally thought to be poorly understood in contrast to some physical systems. For example, we do not understand enough about fish population dynamics and fish behavior to accurately predict strong year classes in a consistent manner. We also do not have a good quantitative idea of trophic energy transfer. In the above cases, we do have some idea of how the fishery systems operate. However, this qualitative knowledge has been difficult to explain and incorporate into management. The English-like knowledge representation in knowledge-based expert systems allows application and testing of this kind of qualitative knowledge.

(b) Expert systems have the ability to handle uncertain information. It is well known that much of the data we gather and parameters we estimate in fisheries investigations are uncertain. That is, there is a considerable amount of sampling error and what we deduce about the future is often dependent upon random events such as the weather. Expert systems employ many methods for dealing with decision making under uncertainty. Examples include Bayes' theory and the Dempster–Shafer theory of evidence. However, at this time there seems to be little consensus about what method is best. Fuzzy logic has also been recently considered.

Expert systems have also received a significant amount of adverse criticism. Some of this relates to the perception of expert systems as only information delivery tools and not research tools. A second and more fundamental criticism has been that there is no evidence that any man-made computer system has demonstrated any reasoning ability or anything resembling human intelligence (Dreyfuss and Dreyfuss, 1986). One response to this is that the inability of an expert system to demonstrate reasoning and intelligence does not detract from the fact that expert systems can aid fishery scientists in utilizing qualitative information and making inferences from it in a systematic and useful manner.

2.2.5 Examples

CLIMPROD, described and developed by Fréon *et al.* (1994), is an interactive software program for choosing and fitting surplus production models, including the consideration of environmental variability. The program contains an embedded expert system to aid in choosing the appropriate model for a given situation and dataset. It also provides assistance in assessing the fit of the model to the data series.

The software is written for PC/XT/AT and higher level microcomputers using MS-DOS version 3.0 or higher. It is a fully interactive program with two main objectives: a normal data management function which utilizes the TURBO C language to provide statistical and graphic capabilities, and a guide for the selection and evaluation of the appropriate model which provides an examination of the way a particular model was chosen. The inference engine for this expert system was written in TURBO PROLOG. Employing backward chaining, it applies about 100 production rules which are interactive with information provided by: (a) questions to the user independent of the data set, (b) statistics about the data set, and (c) graphical deductions from the data set. The program is structured and does not necessarily use the entire range of questions.

CLIMPROD is an excellent example of an expert system embedded within a comprehensive computational model and data. The scope of the domain of the expert system in CLIMPROD is very reasonable, since it only addresses various stock production models and an environmental variable. Broader subject domains, such as the entirety of fish stock assessment and management, are considered to be less appropriate at this time. Indeed, examples from other disciplines bear this out. Two commercially available software packages which include embedded expert systems are Forecast Pro^{TM} and Exsample^{TM}. Forecast Pro^{TM} deals with time-series forecasting methods in which the expert system interacts with the user in choosing an appropriate forecasting model based on the length and nature of the time-series data. Exsample^{TM} guides the user into a choice of appropriate sample size for various types of experiments and analyses, such as analysis of variance and multiple regression. Exsample^{TM} also utilizes an expert system shell Exsys^{TM} to develop the sampling choice protocol.

These are only a very few samples of the expert systems available. Regrettably, there are not yet many in fisheries science.

2.2.6 The future

As the use of knowledge-based systems will continue to expand in the next 5–10 years, fishery scientists will begin to catch up with more established users and developers. I believe that there will be more integration of expert systems with existing databases, management information systems, simulation models, and planning systems. Such systems are called integrated expert systems, and they attempt to compile the data information and knowledge needed to solve problems in a relatively broad domain and make it all accessible for a single computer program. This trend toward integration with other systems – as well as the trend in embedding expert systems into various fisheries application programs – is expected to continue. Expert systems for fisheries will probably become more model based and will capture more of the underlying functional processes of such tasks as classification, diagnosis, and interpretation. There will almost certainly be improved user interfaces through such technologies as multimedia, virtual reality, and improved ways of automating the knowledge acquisition process. It is also relatively easy to imagine a knowledge-based system combined with sensors and automated machinery which will become integral components of a fishing vessel of the future.

The overall advantages of knowledge-based systems are such that progress in their development and implementation seems inevitable. Plant and Stone (1991) describe in detail an integrated decision support system for agriculture which provides a dramatic view of the future of such systems. Such systems are also expected in the future for fisheries management purposes.

2.3 NEURAL NETWORKS

2.3.1 Background

Neural networks, as currently implemented by computer simulations, are collections of relatively simple, highly connected processing elements that respond ('learn') according to given sets of inputs. The basic building block of neural networks technology is the simulated neuron, which processes one or more inputs to produce an output. Individual neurons in a network are sometimes called nodes. The details of how the neurons interconnect are some of the important choices to be made when building a neural network.

The detailed structure of a neural network can be described in terms of the individual neurons, the connections between them, and the activation and transfer functions. These terms seem to be common to many neural networks.

Neurons are located in one of three types of locations: the input layer, the output layer, or one or more hidden layers. Input neurons receive data from the outside world. In general, no processing is done by input neurons. Output neurons are interpreted by the way the neurons are defined in the first place. Hidden neurons are all the neurons in between. Normally the inputs and outputs for hidden neurons are not seen because they connect only to other neurons.

A connection is a unique line of communication that carries a signal from one neuron (a sending neuron) to another (the receiving neuron). Connections can be inhibitory or excitatory. Inhibitory connections prevent firing of a neuron and excitatory ones tend to cause firing. On each connection at the input of a neuron there is a weight, or connection strength, which is analogous to a biological neuron's synapse. The weight controls the strength of the incoming signal to the neuron. The weight of a particular connection may be represented by w_{ij}, where i is the receiving neuron and j is the sending neuron. In matrix notation, one matrix is required for each layer of neurons.

Neurons process inputs and produce outputs. Each neuron takes input from several other neurons. The neuron calculates its output by finding the weighted sum of its inputs. The connection from hidden neuron j's output to output neuron k's input may be represented by w_{kj}, where w is the weight of that connection in the net total effect of the inputs. At any point in time (t), the neuron adds up the weighted inputs to produce an activation value $a_i(t)$ whose value is the sum of weighted inputs at that time. Common activation values are $+1$ and -1, where $+1$ means active and -1 means not active. The activation is passed through an output or transfer function f_i, which produces the actual output for that neuron for that time $o_i(t)$. In the simplest neural network models, the activation function is the weighted sum of the neuron's inputs. In symbolic form,

$$\text{net}_i = \sum_{j=1}^{n} w_{ij} \times o_j$$

That is, the net signal value for neuron i equals the sum of the weight times the input signal for all the inputs to the neuron i from neuron j starting at output of neuron $j = 1$ and ending at $j = n$.

A transfer function of a neuron defines how the activation value is output. A threshold function is an all or nothing transfer function. If the activation value is below the fixed threshold amount, the neuron provides an output of 0. If the neuron activation value is greater than the threshold, the neuron provides an output of 1. Sometimes the transfer function is a saturation function. That is, more excitation beyond some maximum has no further effect. In simple neural networks, a useful transfer function is called the sigmoid function or logistic function:

$$f(x) = \frac{1}{1 + e^{-x}}$$

This function has a high and low saturation limit, and a proportionality range in between.

A neural network is usually trained by either supervised or unsupervised learning. In supervised learning, the more common method, each sample in the training data set, which contains all the inputs as well as the desired output, is presented to the network, and the network 'learns' the patterns presented. In unsupervised learning there is also a collection of sample inputs. However, outputs from the sample set are not provided. The learning process consists of letting the network discover consistent features of the training set, and use these features to group the inputs into clusters that the network finds to be more or less distinct. The Kohonen network (Kohonen, 1988) is an example of a neural network used to discover patterns in data without using other information. Eberhart and Dobbins (1990) provide a very readable introduction to the self-organization neural network model developed by Kohonen. These networks consist of two layers: an input and output layer. The two-dimensional arrays of neurodes in Kohonen networks are called slabs, and they simplify network diagrams by permitting the presentation of groups of neurodes by one symbol. A listing of the source code (written in the C language) for operating a Kohonen network is available in Appendix B of Eberhart and Dobbins (1990).

Normal operation of a supervised learning network is a selective response to a signal pattern. Associated with such a network is a learning rate, which changes the weights. The learning rate is the heart of the neural network, and it is what makes the neural network different from standard computer programs. Normally the learning rate defines how to change the weights in response to summed input/output pairs. Many kinds of learning skills have been developed. Examples include Hebb's rule, the Delta rule, and backpropagation.

2.3.2 Neural network models – backpropagation

Although there are believed to be more than a dozen neural network architectures in general use, this section will be restricted primarily to one, namely, the backpropagation network. Backpropagation was the name given to the first practical method for training multiple-layer feed-forward networks, and it remains a common learning algorithm for such networks. Rumelhart and McClelland (1986) first described the algorithm when they reestablished interest in neural networks. Although Masters (1983) considered backpropagation to be a regrettable misnomer for multiple-layer feed-forward networks, this term seems well established, and it is in common use. A simple three-layer network structure with one output node is shown in Figure 2.2. In this figure, simulated neurons are depicted as circles. The neurons are connected by weights, which are indicated as lines. These weights are applied to values passed from one neuron to the next.

Backpropagation requires training data to provide the network with experience before it is used for processing other data. Externally provided correct patterns are compared with the neural network output during training. Feedback is then used to adjust the weights until all the training patterns are categorized by the network as within the specified tolerance.

For any output neuron j, the error is

$$\text{delta} = (X_j - Y_j) \times f'$$

where X and Y are the desired and the actual outputs, respectively, and f' is the slope of the logistic function (in this example) evaluated at the jth neuron. Then $f' = df/dt = f(1-f)$, where $f(x) = 1/1 + e^{-x}$, and x is proportional to the sum of the weighted inputs of the jth neuron. Other expressions can be used in place of the logistic function to work backward from the output neurons through the inner layer(s) to determine the corrections to their associated weights.

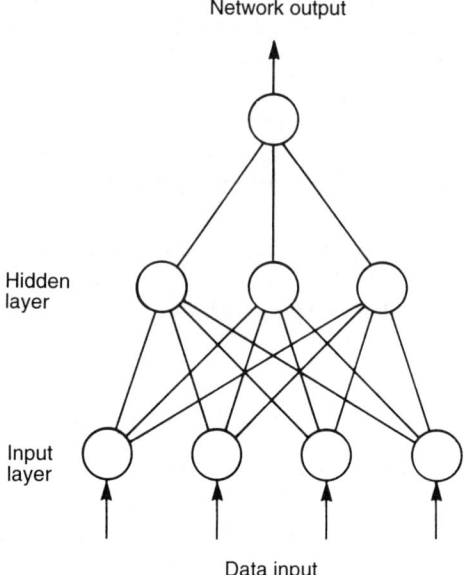

Fig. 2.2 Simple neural network structure, such as might be employed in a three-layer backpropagation network.

The procedure for executing the backpropagation learning algorithm roughly involves the following:

(a) Random values are assigned to the initial weights.
(b) An input vector with the desired output is read in.
(c) Actual output is calculated via the calculations forward through the layers.
(d) Weights are changed by calculating errors backward from the output layer through the hidden layers.
(e) This procedure is repeated for all input vectors until the desired and actual outputs agree within some predetermined tolerance.

The number of nodes (neurons) in the input and output layers are clearly specified by the number of inputs and outputs. However, the number of nodes to use in the middle layers (the hidden layer) is not so clear. Experience has shown that learning improves as the number of hidden nodes increases, but only to a degree. After that, the system performance improves very little, and the learning time increases. There seems to be no single formula which is appropriate for determining the number of hidden nodes to utilize. Only experience and trial and error seem effective.

Two other model parameters, the learning rate and the momentum factor, affect the network's operation. The learning rate (usually termed η) is a constant of proportionality. That is, if the learning rate is set to, say, 0.5, then the weight change is a function of only half the error. The larger the value of η (the learning rate) the greater the weight changes and, therefore, the faster the learning. Large learning rates lead often to convergence, which is not optimal, or to oscillations. One way to avoid oscillation is to make the weight change a function of the previous weight change to provide a smoothing effect. The momentum factor determines the proportion of the last weight change that is added to the new weight change.

The following formula indicates how a particular backpropagation network relates learning rate and momentum to the weight change between two nodes following the presentation of a particular sample case:

new weight change = (learning rate) × (errors) + (momentum) × (last weight change)

The neural net application program illustrated in this section is Neuroshell^{TM}. The above approach, with default learning rate and momentum factors of 0.6 and 0.9, respectively, were employed in the examples to be provided in this section.

For more details concerning the mathematics of neural networks, refer to Rumelhart and McClelland (1986) and to Simpson (1990). White (1989a,b) provides a lucid statement of how some neural network learning procedures are inherently statistical techniques, and how statistical theory can provide insight into the advantages and disadvantages of some neural network learning procedures. Additional information on backpropagation networks is nicely described in Gallant (1994) as well.

2.3.3 Implementation

The number of implementations of neural networks to fishery science-related problems is even more limited than for expert systems. Again, it is evident that ocean engineers have been considerably more active in this area than fishery scientists. However, a brief listing of published papers utilizing neural networks in the broad areas of marine science and fisheries is shown in Table 2.2. This table is not meant to be inclusive. Instead it is intended to provide a sampling of neural network applications for fisheries and marine science.

2.3.4 Forecast comparisons

Many time-series prediction algorithms are already in existence, some of them with solid theoretical foundations and proof of competence. Several of these have been applied by fishery scientists for forecasting. See, for

Table 2.2 Some examples of neural network applications in marine science and fisheries

Reference	Summary
Aoki and Komatsu (1991)	sardine forecasting
Badran *et al.* (1991)	wind ambiguity removal
Balfourt *et al.* (1991)	algal identification
DeSilets *et al.* (1992)	salinity prediction using backpropagation
Frankel *et al.* (1989)	phytoplankton flow cytometric data analysis
Ghosh *et al.* (1992)	oceanic signal detection, classification, and characterization
Maccato and deFigueiredo (1989)	classification of ocean acoustic signals
Moore *et al.* (1989)	dolphin echo location
Potter *et al.* (1991)	salmon scale classification
Silven (1992)	multiple target tracking
Simpson *et al.* (1993)	classification of *Dinophyceae*
Smits *et al.* (1992)	algal classification

example, Saila *et al.* (1979), Mendelsohn (1981), Jensen (1985), and Tsai and Chai (1992). Why then consider alternatives? Masters (1993) indicates that there are at least two reasons for considering a neural network alternative to more traditional forecasting techniques:

(a) Neural networks are sufficiently versatile that they do not require choosing a particular model from the many alternatives. As is well known by users, considerable judgment is involved in choosing an appropriate traditional time-series model.
(b) Strange noise patterns, including chaotic components, are tolerated by neural networks far better than by traditional methods.

For example, Lacepedes and Farber (1987) demonstrated that neural networks can achieve significantly higher numerical accuracy than more conventional forecasting techniques in predicting future values of a highly chaotic time series. However, a time series for a very large number of steps (typically 500) was utilized in their study, and this is considerably more than anticipated in typical fishery applications. It appears from the above study that the most difficult systems to forecast are those that are nonlinear and chaotic. A truly chaotic series was not considered herein, but in order to produce a synthetic data set containing both signal and noise (believed to be typical of much fishery data) which could be compared by the various forecasting techniques, a 50-period time series of data was generated from a model described by Berryman *et al.* (1986) and Berryman and Millstein (1990). This time series was generated from the following equation which incorporates both density-dependent and density-independent factors:

$$R = A\left[1 - \frac{N(t-1)}{L}\right]\left[1 - \left(\frac{N(t-T)}{K}\right)^{Q}\right] + V(S)$$

where R is the per individual rate of increase of the population. R is also a function of its own population density plus the random effect of

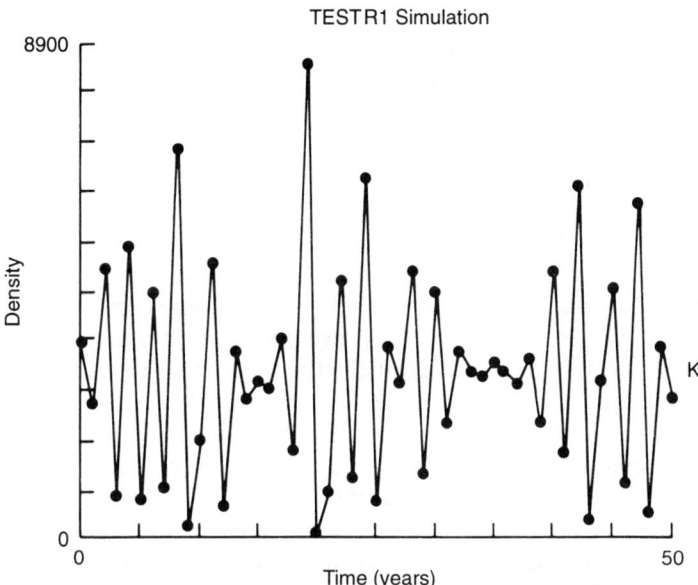

Fig. 2.3 Plot of TESTR1 data generated by the generalized logistic model and the parameters given in the text. Since the random seed number was not retained, the entire data set is listed to four significant digits. Only the first 40 data points were used in the development of all models tested. Model parameters were:

$$
\begin{array}{lcl}
A & = & 2.5 \\
K & = & 3000 \\
L & = & -1000000 \\
Q & = & 1 \\
T & = & 1 \\
S & = & .25
\end{array}
$$

Data point values:

(1) 2422	(11) 4944	(21) 887	(31) 2044	(41) 1510
(2) 4884	(12) 6404	(22) 4675	(32) 3380	(42) 6276
(3) 7599	(13) 3369	(23) 1119	(33) 3002	(43) 366
(4) 5208	(14) 2494	(24) 6468	(34) 2873	(44) 2287
(5) 7381	(15) 2817	(25) 679	(35) 3138	(45) 4461
(6) 4419	(16) 2658	(26) 3461	(36) 2980	(46) 1009
(7) 9125	(17) 3576	(27) 2791	(37) 2723	(47) 5974
(8) 7016	(18) 1588	(28) 4808	(38) 3170	(48) 488
(9) 2534	(19) 8522	(29) 1166	(39) 2064	(49) 3463
(10) 1794	(20) 109	(30) 4398	(40) 4819	(50) 2554

environmental factors such that $R = f[N(t - T)] + V$, where f is the density-dependent R function, N is the current population density, T is the delay in the feedback response, and V is a random variable with mean $= 0$ and standard deviation $= S$. Also, A is the value of R at $N = 0$, L is the low-density root or extinction threshold, K is the carrying capacity of the environment, and Q is the coefficient of curvature of the density-dependent R function.

For the time series illustrated in Figure 2.3, which was generated by this model, the parameters were set at: $A = 2.5$, $L = -1.00 \times 10^6$, $K = 3000$, $Q = 1.00$, $T = 1.00$, N at $T_0 = 3500$, and $S = 0.25$. The actual values for each data point are also given because the seed for the random number generator was not listed, and only approximations to these data can be generated due to the random element. The data are given in the hope that others may wish to compare their forecast methods with those presented herein or with the results obtained. The software used for fitting the Box–Jenkins and dynamic regression or models were Forecast Pro™ and Forecast Master™. The neural network (NN) backpropagation algorithm was from Neuroshell™. Although the author had initially written some software for forecasting, the use of commercial software was thought to be helpful in permitting replication of the results obtained, as well as permitting easy access to some neural network algorithms. However, it should be pointed out that there is no guarantee that the results obtained from application of the backpropagation neural network algorithm converge to a global minimum, or that someone using the same software will achieve exactly the same results due to the variations in the convergence cutoff point and other adjustable parameters such as momentum and learning rate.

For the development of all models shown in Table 2.3, only the first 40 data points were utilized. The next five points were compared with the

Table 2.3 Model prediction comparisons for TESTR1 data shown in Figure 2.3. Note that only the first 40 points were used in developing these models and the predictions made from them. Time steps 1–5 refer to data points 41–45.

		Model predictions		
Time steps	*Real data*	*Box–Jenkins ARIMA (0,0,1)*	*Dynamic regression 3 lagged variables*	*Neural network 3 lagged variables*
1	15.102	28.802	22.868	16.143
2	62.764	31.472	33.797	80.217
3	3.659	31.472	26.936	5.962
4	27.868	31.472	35.934	30.574
5	44.613	31.472	28.530	70.092

Table 2.4 Statistical comparisons of the predictive models using TESTR1 data shown in Table 2.3. The prediction statistics are described in Power (1993).

	Prediction model		
Prediction statistic	Box–Jenkins	Dynamic regression	Neural network
Mean error	0.137	−1.188	13.796
Mean percent error	156.869	135.730	32.882
Mean square error	426.223	352.527	193.499
Mean absolute error	17.910	16.832	9.796
Thiel's U^2	0.306	0.254	0.139

various model predictions. It seems clear from the results shown in Table 2.3 that the neural network provided substantially more accurate forecasts of the real data except for the fourth-step-ahead forecast from the Box–Jenkins model. This was believed to result from the fortuitous coincidence of the real data and the extension of an identical prediction from the second forecast. A statistical comparison of the model predictions for the TESTR1 data is shown in Table 2.4. Only the mean error estimate (bias) is higher in the NN model. The relatively high mean percent errors in the traditional model predictions seemed to be better balanced between negative and positive deviations from actual values, which explains the lower mean error estimate. However, mean square error and absolute error were smaller for the NN model. The value of Thiel's statistic suggests that all three models have some predictive power but that the NN model coefficient was closer to zero, which is indicative of higher predictive ability.

The next data set used for comparative purposes was reported by Tsai and Chai (1992). They used a time series of Maryland juvenile striped bass indices and commercial landings from Maryland (1954–1984) to compare a classical regression model, univariate ARIMA model, a time-series regression model, and a transfer function noise model in terms of fitting errors and prediction errors for one-step-ahead forecasts. Their major conclusion was that the forecast error from two models, namely the transfer function noise model and the time-series regression, provided the best fits to the data, but that the mean absolute forecast errors of 84.56 and 80.74 percent, respectively, were too high for effective practical forecasting. It seemed desirable to compare the results obtained from an NN model with the outputs from these models. Therefore, the same data was used and output comparisons were made. Table 2 of Tsai and Chai (1992) was extended to form Table 2.5 of this report. The NN model was superior to

Table 2.5 Forecasts (F) and corresponding forecast errors (FE = forecast − actual harvest, RMSE$_f$ = root mean square forecast errors, MAPE = mean absolute percent forecast error) of the classic regression model, univariate ARIMA model, time series regression model, transfer function noise model, and neural network model for the striped bass commercial harvests in the Maryland portion of Chesapeake Bay, 1981–1984.

Forecast year	Actual harvest (t)	Classical regression model		Univariate ARIMA model		Time series regression model		Transfer function noise model		Neural network model	
		F (t)	FE (t)	F (t)	FE (t)	F (t)	FE (t)	F (t)	FE (t)	F (t)	(FE (t))
1981	896.87	1665.30	768.43	1121.90	225.50	1286.41	389.54	1246.80	349.93	977.0	80.1
1982	242.81	1423.16	1180.35	896.87	654.06	690.23	447.42	737.59	494.78	224.6	18.2
1983	238.02	894.86	656.84	242.81	4.76	326.36	124.34	247.44	9.42	230.9	38.0
1984	724.03	1002.70	278.67	238.02	−486.01	412.10	−311.43	348.24	−375.75	200.0	524.0
RMSE$_f$(t)			789.43		422.75		340.72		356.56		265.9
MAPE$_f$(%)			184.51		90.90		80.74		84.56		26.2

the traditional models with respect to simulation errors (root mean square and absolute errors), but it was not as accurate nor precise with respect to forecasts and forecast errors when the entire data set (1954–1980) was used in model development. However, examination of both landings, as well as juvenile index data illustrated in Tsai and Chai (1992), clearly demonstrated some change or breakpoint in the basic process states starting at or about 1970. This change in basic process states was previously recognized by Tsai *et al.* (1991), and they even provided a hypothesis regarding causality. Harrison and Stevens (1971) warned that the performance of traditional forecasting models could be adversely affected by changes in trend or slope of the data. Although Tsai and Chai (1992) attempted to account for trends by differencing, their models (especially the Box–Jenkins procedures) could not accommodate small data sets effectively. On the other hand, the neural network model is more versatile, and it was possible to utilize data only from the 1970–1980 period to generate the forecasts shown in Table 2.3. Specifically, this limited data set was detrended using linear regression and scaled. Standardized residuals were used as inputs to the neural network model. Both landings and the juvenile index were lagged for two time periods to provide four inputs to the neural network. Then NN model predictions were more accurate than those from any model using the entire data set. Only the last year (1984) had an unacceptably high forecast error. Despite this, the root mean square error and the mean absolute percent forecast error were lower than those achieved by all other models.

Another form of comparative analysis involved assessing the relative predictive values of the standard stock-assessment models (Ricker, 1954; Beverton and Holt, 1957) with NN forecasts using the same data set. Power (1993) had already provided model predictions and statistical comparisons of the above mentioned two models utilizing stock-recruit data for brown trout reported initially by Elliott (1984). The data from Tables 2 and 3 of Power (1993) were extended to provide results using the NN model to compare with the two stock-recruit equation projections. From Table 2.6 it is clear that the NN model predicted accurately in the first year, less well in the second, and better in the third. The fourth was slightly higher for the NN model, and the fifth was the same for the three models. The statistical comparison of the predictive models (Table 2.7) illustrates that the mean absolute error and mean absolute percent error of the NN model were smallest of the three model types, but the mean error, mean percent error, and mean square error were intermediate. However, the values of Thiel's U^2-statistic are greater than 1 for both the Ricker and Beverton–Holt (B–H) models, indicating that the predictive power of these models is no more than a naive no-change prediction. The Janus coefficient (J^2) provides an index of the change in model accuracy between the estimation and predic-

software programming, it seems likely that combining neural networks, expert systems, database and conventional software into powerful hybrid systems will further enhance the development of optimal solutions to complex fishery-related problems. Indeed, it seems possible that intelligent systems may eventually mimic human decision making under uncertainty, even in situations where information is incomplete or contains errors.

Another interesting area for future fisheries research is the application of chaos theory to address complex system stability or system change. Chaos research attempts to determine how complex systems such as ocean currents and fish reproductive success are organized, and how small changes in some inputs can produce profound changes in the system, resulting in chaotic behavior. As suggested previously, recurrent neural networks are able to represent such systems.

The potential of neural networks for fisheries implementation is believed to be high, and the future should be truly exciting.

2.4 GENETIC ALGORITHMS

2.4.1 Background

Although classic genetic algorithms (GAs) were first developed by Holland (1975) and later extended by Goldberg (1989) and others, the application of genetic algorithms to fishery science and related problems has been extremely limited to date. This is unfortunate because certain problems in fisheries involve large and complicated search spaces for their solution, and genetic algorithms are adaptive search techniques which tend to efficiently converge on solutions that are globally optimum or nearly so without exhaustive search. Many practical problems in fisheries have involved finding an optimal solution by sampling through the entire set of possible solutions. Such an exhaustive search is sometimes described as exploring the 'total parameter space' of the problem, and it can be very time consuming even with powerful computers. Therefore, the usual approach has been to iteratively refine a trial solution until no further improvements within defined limits are found. Genetic algorithms operate differently to conventional optimization procedures because they are based on reasonable analogies to biological evolution – hence the name. In a sense they 'breed' trial solutions to a problem, and only the best solutions survive after several generations. Austin (1990) provides a very brief and also very readable introduction to genetic algorithms, which also includes a description of some applications. A good example of the specific application of a genetic algorithm to a computationally difficult scheduling problem is provided by Lawton (1992). DeJong (1988) also provides an overview of learning with genetic algorithms.

2.4.2 Description

An excellent introduction to genetic algorithms with C++ is provided by Singleton (1994). It not only quickly explains how GAs work but also briefly indicates how one can use a genetic algorithm to generate computer programs. Neuroshell 2, Release 2.0, already uses a GA to find smoothing factors for certain neural networks. The following is a very simplistic attempt to briefly describe GAs; basically, the principle behind them is natural selection. From an evolutionary point of view, each species is searching for beneficial adaptations to a complicated and changing environment. The 'information' each species has gained is embodied in the makeup of its chromosomes. Chromosomal makeup is modified when reproduction takes place. Chromosomal modifications include: random mutations, inversion of chromosomal material, and crossover-exchange of chromosomal material between two parents' chromosomes. Crossover seems to be a key element in the effectiveness of genetic algorithms.

According to Davis and Steenstrup (1987), a genetic algorithm must have five components in order to solve a problem:

(a) a chromosomal representation of solutions to the problem;
(b) a way to create an initial population of solutions;
(c) an evaluation function that plays the role of the environment;
(d) genetic operators that alter the composition of 'offspring' during reproduction;
(e) values for the parameters that the genetic algorithm uses (population size, probabilities of applying genetic operators, etc.).

A great deal of study has been devoted to each of these five components of genetic algorithms. No effort is made to summarize these herein; Davis and Steenstrup (1987) provide more details.

Basically, a genetic algorithm is a stochastic search algorithm with the following properties:

(a) It represents possible solutions as strings (chromosomes).
(b) It creates generations of solutions using operators from genetics, such as crossing over and mutations to create new variations in the population.
(c) It evaluates each individual, based on fitness criteria related to a specific problem (the environment).
(d) It can use the fitness information to favor the best individuals (strings) when creating the next generation.

In GAs the environment is simulated using appropriate constraints and suitability functions (fitness functions). The results of the 'evolutionary' process are achieved by successive improvement from generation to generation.

2.4.3 Implementation

Problems to be solved with GAs need to be posed in such a way that allows solutions to be described as strings of numbers or characters. Each symbol can represent one or more complex operations. A string, such as ACBDF, might represent the order in which the solution to a problem is implemented. This string is called a chromosome. A particular solution has certain properties or consequences that give it a relative worth as a solution to a particular problem. An evaluation function produces a numerical value that represents the 'chromosome's' ability to solve the problem. The GA uses the genetic operators of reproduction, mutations, and crossover for searching for an efficient optimum or near optimum solution.

There are several software packages already available for users interested in applying genetic algorithms, a few which are mentioned below.

Evolver™ is a software package designed to find optimal solutions to some classes of problems. Version 2 of Evolver™ requires Microsoft Excel 3.0 or 4.0 running under the Windows™ operating system. It is essentially an add-in to Excel™, which first involves building a system model in Excel™, although advanced users can build customized, stand-alone programs. The Evolver™ package uses a proprietary set of genetic algorithms to find optimum solutions to a problem. The Evolver's options are for difficult nonlinear problems in contrast to the solver section of Excel™, which addresses less complex optimization problems.

MicroGA™ is another implementation of the GA method of problem solving. It requires a PC-compatible computer with a 286 or higher processor, Borland C++ and Microsoft Windows™ 3.0 or higher. This software package is written using Borland C++ Version 3.1 for Windows™. MicroGA™ is a framework of C++ objects. A framework is different from a class library. The framework is designed so that several pieces are used in conjunction with each other in order to give the user some default behavior.

CDARWIN II™ is an optimizing PC tool based on genetic algorithms. The program requires an IBM PC or compatible computer with MS-DOS 3.2 or higher and 640 KB of RAM. The basic steps in solving a problem with CDARWIN II are:

(a) define the problem and select a suitable representation;
(b) select and define the chromosome and genes;
(c) select the data and enter functions and constraints;
(d) run the program and follow convergence.

The most promising development areas for genetic algorithms in fisheries appear to lie in the adaptive control of complex processes. In general, conventional control theory seems well suited for applications wherein the process can be reasonably well characterized in advance and where the number of parameters to be considered is small. Fisheries involve a class of processes that are not well characterized and are subject to a number of uncontrolled or highly variable parameters. Genetic learning appears to be a feasible means for achieving good performance control rules without having good models of the process being controlled. Although the system for learning control strategies with genetic algorithms was illustrated with aircraft evasive maneuvers, the learning system described by Grefenstette (1989) seems suitable for rule-based control of fisheries with relatively little modification.

Although unable to find any fisheries or marine science-related references to genetic algorithms in a search of the most recent version of ASFIS, some related material was found. An example includes the work of Perry (1984) on speciation in ecological niche theory using genetic algorithms. In addition, a genetic algorithm has been applied by the author to the optimum routing of fishing vessels, which is an application of the genetic algorithm to the so-called traveling salesman problem.

Other interesting developments with genetic algorithms include using the genetic algorithm to optimize neural network architectures (Schaffer *et al.*, 1990). As indicated previously, a difficult task in developing a neural network is finding the optimum tradeoff among several adjustable parameters, such as number of nodes, hidden layers, and best starting weights and learning rate. Reitman (1994) provides two genetic algorithms for neural networks.

As a final intriguing thought, it has been demonstrated that artificial evolution as described by Ackley and Littmen (1992) can result in diverse species with genome distributions resembling biological populations in the real world.

2.5 SIMULATED ANNEALING

Simulated annealing has been described by Masters (1993) as an effective means for avoiding local minima, as well as for escaping from them, if necessary, in error surfaces encountered in neural network learning. Masters (1993) has also provided a general optimization subroutine using simulated annealing, which is written in C++. Much of what follows is a very brief review of Chapter 7 in the above reference.

Annealing is a term which has generally been applied to a physical process, especially as related to metallurgy. A metal tends to be rather brittle when the atoms in the piece of metal are arranged randomly. In the annealing process, the metal is heated to a very high temperature, which induces the atoms in the material to oscillate relatively violently. If the metal is cooled very quickly, the atoms tend to become locked in a random state. On the other hand, if the metal is cooled very slowly, the atoms tend to fall into patterns which are quite stable for that temperature. If the cooling is sufficiently slow, the metal ultimately stabilizes in an orderly atomic structure which is stronger than that found in the random arrangement of the atoms.

Simulated annealing, as applied to an optimization process, involves the random perturbation of the independent variables (connection weights in a neural network) and keeping track of the best (lowest error) function value for each randomized set of variables. A relatively high standard deviation (i.e. simulated annealing temperature) is used at first. After many trials, the set of variables which produced the lowest error is designated the center about which perturbations will take place for the next temperature. The standard deviation (temperature) is then reduced, and new trials are done. No details are provided herein regarding the choice of the rate of reduction of the standard deviation nor of the number of trials required.

It is interesting to note that simulated annealing is both an artificial intelligence tool and a stochastic computational technique. However, it is based on a physical process in contrast to the biological processes which form the basis of both neural networks and genetic algorithms. The theoretical foundations of the annealing process are found in statistical mechanics. Davis and Streenstrup (1987) provide some introduction to simulated annealing and its applications. In general, the simulation of annealing applied to optimization problems involves the following:

(a) Identify the analogs of the physical concepts in the optimization problem. For example, the energy function becomes the objective function; the configuration of particles becomes the configuration of the parameter values; finding a low-energy configuration becomes seeking a near-optimal solution; and the temperature becomes the control parameter of the process.
(b) Select an annealing schedule consisting of decreasing temperatures and the time to spend at each temperature.
(c) Develop a means for generating and selecting new configurations.

In general, finding the global minimum of an objective function with many degrees of freedom subject to conflicting constraints is termed an NP-complete problem, since the objective function will tend to have many local minima. A suitable procedure for solving hard optimization problems should sample values of the objective function in such a way as to have a high probability of finding a near-optimal solution and lend itself to efficient implementation.

This section has treated simulated annealing in a very cursory fashion. However, the role of simulated annealing in the solution of tough optimization problems should not be underestimated, and fishery science stands to benefit from these developments.

2.6 SOME FUTURE DIRECTIONS FOR ARTIFICIAL INTELLIGENCE

2.6.1 Background

This summary of future directions for artificial intelligence implementations in fishery science is based largely on the projections of experts in the various AI disciplines, gleaned from the literature. These future developments are now placed in a slightly more fisheries-related context.

2.6.2 Expert systems

It is clear that there will be an expansion of interest and activity regarding expert system implementations in fishery science. As pointed out by Holtzman (1989), steps are already being taken to develop 'intelligent decision systems'. These systems combine the knowledge-based reasoning methods of expert systems with formal methods of decision analysis. Holtzman proposes combining the techniques of the knowledge engineer and the decision analyst by applying knowledge engineering to the tasks of the decision analyst. The effective decision system of the future will concentrate on assisting the decision maker (manager) to give insight into the decision procedures at hand rather than merely supplying a somehow 'right' answer. This implies that the decision system must concentrate on formulating a problem clearly and understandably rather than focusing exclusively on problem solving. The challenging task for the near future will be how to effectively represent and use uncertain knowledge, because uncertainty requires us to represent and reason about a multiplicity of possible scenarios. As the size of the problem grows, the number of ways in which uncertainty can be resolved becomes extremely large.

It seems evident that future knowledge-based systems in fisheries will be integrated with database management, prediction systems, automated sensors, and planning systems. It is conceivable that in the future, fishing activities and fishery management will be largely integrated. Telemetry from monitoring buoys will be utilized tactically by the fishermen and for strategic planning by the fishery managers. In this scenario, simulation models are run by the manager to provide resource assessment information, while the fishermen will attempt to optimize the capture activity within the constraints provided by the manager.

2.6.3 Neural networks

There is no question that neural networks are here to stay, and that the most exciting conceptual developments and applications are yet to come. Their future in fisheries science seems particularly promising.

According to Eberhart and Dobbins (1990) there are at least four application areas for which neural networks are considered to be best suited. There is much room for future development of fishery science implementations in each of the following areas:

(a) Classification as reflected in decision theory – this refers to deciding which of several predefined classes best reflects an input pattern. One example from fisheries is a decision whether or not a given individual fish represents a particular stock.

(b) Content addressable memory (associative memory) – an example of this is obtaining the complete version of a pattern as the output of a network by providing a partial version at the input.

(c) Clustering (compression) – an example is reducing the number of bits that must be stored or transmitted to represent a block of data within a specified error.

(d) Generation of structured sequences or patterns from a network trained to examples – an example is a neural network trained to model or simulate a system. Because of inherent randomness in the processes being simulated, there may be no 'correct' answers. However, the system must be described statistically, and a neural network may be designed to reproduce these qualities.

This number of applications of neural networks to the above-mentioned four areas will surely grow in the near future. Indeed, it is anticipated that the number of areas for neural network application will also expand in the future. Fishery scientists should find neural networks a powerful tool for further development.

2.6.4 Genetic algorithms

Genetic algorithms have a very impressive record of progress as a heuristic search method. Many practical problems in engineering and science require a search through the entire set of all possible solutions in order to find an optimal solution. The process of genetic algorithms derives from their emulation of nature's principles of evolution over generations of a species. In the case of searches for optima, a population of candidate solutions is allowed to evolve over several generations, with the fittest individuals having the best chance for survival.

Efficient search features and wide applicability are the attractive properties of genetic algorithms, which will make them important tools to the fishery scientist of the future.

Finally, GAs will be utilized to optimize neural network outputs, as well as initial weights and learning rates.

REFERENCES

Ackley, D. and Littmen, M. (1992) Interaction between learning and evolution, in *Artificial Life II*, (eds J.D. Farmer and S. Resinussen), Addison-Wesley, Redwood City.

Aoki, I., Inagaki, T., Mitani, I. and Ishii, T. (1989) A prototype expert system for predicting fishing conditions of anchovy off the coast of Kanagawa Prefecture. *Bulletin of the Japanese Society for Scientific Fisheries*, **55**(10), 1777–83.

Aoki, I. and Komatsu, T. (1991) Neuro-computing for forecasting the catch of young sardine. *Bulletin Japanese Society for Fisheries and Oceanography*, **56**(2), 113–20.

Austin, S. (1990) An introduction to genetic algorithms. *AI Expert*, **5**(3), 49–53.

Badran, F., Thiria, S. and Crepon, M. (1991) Wind ambiguity removal by the use of neural network techniques. *Journal of Geophysical Research*, **96**(11), 521–9.

Balfourt, H.W., Snoek, J., Smits, J.R.M., Breedveld, L.W., Hofstraat, J.W.H. and Ringelberg, J. (1991) Automatic identification of algae: Neural network analysis of flow cytometric data. *Journal of Plankton Research*, **14**(4), 575–89.

Bender, M.J., Ketopodis, C. and Simonvic, S.P. (1992) A prototype expert system for fishway design, in *Expert Systems and Statistical Methods in Water Resources*, (eds A.H. Elshaarwi and D.C.L. Lam), Vol. 23 (1–5), 115–27.

Berryman, A.A. and Millstein, J.A. (1990) Population Analysis System, Version 2.5. Ecological Systems Analysis, Pullman, Washington.

Berryman, A.A., Stenseth, N.C. and Isaev, A.S. (1986) Natural regulation of insect herbivores in forest ecosystems. *Ecologia*, **71**, 174–84.

Beverton, R.J.H. and Holt, S.J. (1957) *On the Dynamics of Exploited Fish Populations*, Fisheries Investigations, London, Series 2, No. 19, 1–533.

Bossu, T., Mantoni, M. and Saroglia, M. (1989) Expert system for sanitary control in intensive fish culture: The case of sea bass reared in thermal effluents, *Aquaculture-Europe '89*, Short Communications and Abstracts of Review Papers, Films, Slide Shows, and Poster Papers, presented at the International Aquaculture Conference held in Bordeaux, France, 2–4 October 1989, pp. 39–40.

Calisal, S.M., Mikkelsen, J., McGreer, D., Akinturk, A. and Havens, W. (1992) Fishing vessel design and constraint reasoning system, in *Design of Marine and Offshore Structures*, (eds T.K.S. Murthy and J.A. Alaez), Elsevier/Applied Science (UK), pp. 183–99.

Davis, J.R. and Clark, J.L. (1989) A selective bibliography of expert systems in natural resource management. *AI Applications in Natural Resource Management*, **3**(3), 1–18.

Davis, L. and Steenstrup, M. (1987) Genetic algorithms and simulated annealing: An overview, in *Genetic Algorithms and Simulated Annealing*, (ed. L. Davis), Pitman, London, 1–11.

DeJong, K. (1988) Learning with genetic algorithms. An overview. *Machine Learning*, **3**, 121–38.

DeSilets, L., Golden, B., Wang, Q. and Kumer, R. (1992) Predicting salinity in the

Chesapeake Bay using backpropagation. *Computers in Operations Research*, **19**(3–4), 277–85.

Dillon, T.M. and Lutz, C.-H. (1991) A computer-assisted expert system for interpreting the consequences of bioaccumulation in aquatic animals (COBIAA). *Environmental Effects of Dredging*, **D-81-4**, 1–5.

Dreyfuss, H.L. and Dreyfuss, S.E. (1986) *Mind over Machine: The Power of Human Intuition and Expertise in the Area of the Computers*, Free Press, New York.

Eberhart, R.C. and Dobbins, R.W. (eds) (1990) *Neural Network PC Tools: A Practical Guide*, Academic Press, New York.

Elliott, J.M. (1984) Numerical changes and population regulation in young migrating trout *Salmo trutta* in a lake district stream, 1966–83. *Journal of Animal Ecology*, **53**, 327–50.

Frankel, D., Olsen, R.J., Frankel, S.L. and Chisholm, S.W. (1989) Use of a neural network computer system for analysis of flow cytometric data of phytoplankton populations, in *Cytometry in Aquatic Sciences* Vol. 10 (eds C.M. Yentsch and P.K. Heren), (5), 540–50.

Fréon, P., Mullen, C. and Pinchon, G. (1994) CLIMPROD experimental interactive software for choosing and fitting surplus production models including environmental variables on IBM PC and compatibles, FAO Fisheries Technical Paper, Food and Agriculture Organization, Rome, Italy (in press).

Froese, R. and Schoefer, W. (1987) Computer-aided identification of fish larvae, Instructional Council for Exploration of the Sea, The Philippines CM-1987/L:23.

Froese, R., Schoefer, W., Roepke, D., Piatkowski, W. and Schneck, D. (1989) Computed-aided Identification of Fish Larvae. *Fishbyte*, **7**(2), 18–19.

Fuchs, F. (1991) An expert system for analyzing relationships between fish and the environment, International Council for the Exploration of the Sea, Copenhagen, Denmark, International Council for Exploration of the Sea, C.M. 1991/D:5, 6 pp.

Gallant, S.I. (1994) *Neural Network Learning and Expert Systems*, MIT Press, Cambridge, Massachusetts.

Ghosh, J., Deuser, L.M. and Beck, S.D. (1992) A neural network based hybrid system for detection, characterization, and classification of short-duration oceanic signals. *IEEE Journal of Ocean Engineering*, **17**(4), 351–53.

Goldberg, D.E. (1989) *Genetic Algorithms in Search, Optimization, and Machine Learning*, Addison-Wesley, Reading, Massachusetts.

Grefenstette, J.J. (1989) A system for learning control strategies with genetic algorithms, in *Proceedings of the Third International Conference on Genetic Algorithms*, (ed. J.D. Schaffer), Morgan Kanfram, San Mateo, California, pp. 83–90.

Hanfman, D., Bielawski, L. and Lewand, R. (1989) REGIS: Regional information system for African aquaculture, National Agricultural Library, Beltsville, Maryland, USA (NTIS Order No. PB89-170328/GAR).

Harrison, P.J. and Stevens, C.F. (1971) A Bayesian approach to short-term forecasting. *Operations Research Quarterly*, **22**(4), 341–53.

Holland, J.H. (1975) *Adaptation in Natural and Artificial Systems*, University of Michigan Press, Ann Arbor, Michigan.

Holtzman, S. (1989) *Intelligent Decision Systems*, Addison-Wesley, Reading, Massachusetts.

Jensen, A.L. (1985) Time-series analysis and the forecasting of menhaden catch and CPUE. *North American Journal of Fisheries Management*, **5**, 78–85.

Kohonen, T. (1988) *Self-organization and Associative Memory*, Springer-Verlag, New York.

Kosko, B. (1992) *Neural Networks and Fuzzy Systems*. Prentice Hall, New Jersey.

Lacepedes, A. and Farber, R. (1987) Nonlinear signal processing using neural networks: Prediction and system modeling, LA-UR-F7-2662, Los Alamos National Laboratory, Los Alamos, New Mexico.

Lawton, G. (1992) Genetic algorithms for schedule optimization. *AI Expert*, **7**, 40–6.

Lee, H., Winsor, M., Pelletier, J., Randall, R. and Bartling, J. (1991) Computerized risk and bioaccumulation system (Version 1.0), Ecological Research Series, US Environmental Protection Agency, 27 pp.

Lehner, P.E. and S.W. Barth (1985) Expert systems on microcomputers, in *Applications in Artificial Intelligence*, (ed. J. Andriole), Petrocelli Books, Princeton, New Jersey.

Maccato, A. and deFigueiredo, R.J.P. (1989) *A neural network based framework for classification of oceanic acoustic signals. Oceans 89: The Global Ocean*, Institute of Electrical and Electronic Engineers, New York, Vol. 4, 1118–23.

Markert, B. and Klausmeuer, N. (1990) Variations in the elemental composition of plants and computer-aided sampling in ecosystems. *Toxicology and Environmental Chemistry*, **25**(4), 201–12.

Masters, T. (1993) *Practical Neural Network Recipes on C^{++}*, Academic Press, San Diego, California.

Mendelsohn, R. (1981) Using Box–Jenkins models to forecast fishery dynamics: identification, estimation, and checking. *Fishery Bulletin*, **78**, 887–96.

Moore, P.W., Nachtigall, P.E., Penner, R.H., Au, W.W. and Roitblat, H.L. (1989) Dolphin echo location: identification of returning echoes using a counter-propagation network, Naval Ocean Systems Center, San Diego, California, USA (NTIS Order No. AD-A211, 805/7/GAR).

Perry, Z.A. (1994) *Experimental study of speciation in ecological niche theory using genetic algorithms*, PhD thesis, University of Michigan, Dissertation Abstracts International 45(12), 387-D, University Microfilms, No. 8502912.

Plant, R.E. and Stone, N.D. (1991) *Knowledge-based Systems in Agriculture*, McGraw Hill, New York.

Potter, E.C.E., Kell, L. and Reddin, D.G. (1991) The use of a neural network to distinguish North American and European salmon (*Salmo salar* L.) using scale characteristics, ICES, C.M. 1991/M:10 pp.

Power, M. (1993) The predictive validation of ecological and environmental models. *Ecological Modelling*, **68**, 33–50.

Rallet, F., Garnerin, P. and Tuffery, G. (1990) Schubert audit: Expert system for assistance in the quality audit of farms producing trout for consumption and re-stocking. First operational mock-up. *Bulletin Francois Peche et Pisceculture*, **316**, 1–14.

Rauch-Hindin, W.B. (1988) *A Guide to Commercial Artificial Intelligence*, Prentice Hall, New Jersey.

Recknagel, F., Beuschold, E. and Petersohn, W. (1991) DELAQUA: A prototype expert system for operational control and management of lake water quality, in *WATERMATEX-91, Proceedings of the Second International Conference on Systems Analysis in Water Quality Management*, Durham, New Hampshire, USA, 3–6 June 1991, (eds T.O. Barnwell, P.J. Ossenbruggen and M.B. Beck) **24**(6), pp. 283–90.

Reitman, E. (1994) *Genesis Redux*, Windcrest/McGraw Hill, New York.

Ricker, W.W. (1954) Stock and recruitment. *Journal of the Fisheries Research Board of Canada*, **11**, 769–99.

Rumelhart, D. and McClelland, J. (1986) *Parallel Distributed Processing*, MIT Press, Cambridge, Massachusetts.

Ryan, J.D. and Smith, P.E. (1985) An 'expert system' for fisheries management, *Oceans '85 Proceedings*, Ocean Engineering and the Environment, Marine Technology Society, Vol. 2, pp. 1114–17.

Saila, S.B., Wigbout, M. and Lermit, R.T. (1979) Comparison of some time-series models for the analysis of fisheries data. *Journal du Conseil*, **39**(1), 49–52.

Schaffer, J.D., Caruana, R.A. and Eghelman, L.J. (1990) Using genetic search to exploit the emergent behavior of neural networks. *Physica*, **D42**, 244–8.

Silven, S. (1992) A neural approach to the assessment algorithm for multiple target tracking. *IEEE Journal of Ocean Engineering*, **17**(4), 326–32.

Simpson, P. (1990) *Artificial Neural Systems: Foundations, Paradigms, Applications, and Implementation*, McGraw Hill, New York.

Simpson, R., Culverhouse, P.F., Williams, R. and Ellis, R. (1993) Classification of Dinophyceae by Artificial Neural Networks, in *Toxic Phytoplankton Blooms in the Sea*, (eds T.J. Smayda and Y. Shimizu), Elsevier, Amsterdam, Vol. 3, 183–90.

Singleton, A. (1994) *Genetic programming with C++*. *Byte*, February, 171–6.

Smits, J.R.M., Breedveld, L.W., Derksen, M.W.J., Kateman, G., Balfourt, H.W., Snoek, J. and Hofstraat, J.W. (1992) Pattern classification with artificial neural networks: Classification of algae, based on flow cytometer data. *Analytical Chimica Acta*, **258**(1), 11–25.

Stagg, C.M. (1990) The expert support system as a tool in fishery stock assessment and management, in *Proceedings of the NATO Advanced Study Institute on Operations Research and Management in Fishing*, (ed. A. Guionares Rodrigues), Povoa de Varzim, Portugal, pp. 299–314.

Stokoe, P.K. and Gray, A.G. (1990) AQUASITE: A computer site assessment system for marine coastal aquaculture, Aquaculture Association of Canada Conference, Halifax, Nova Scotia, **90**(4), pp. 94–6.

Stretta, J.M., Petit, N. and Simier, M. (1990) Integration of aerospatial remote sensing into the data base associated with an expert system for the tuna fishery, in *French Contributions in the French–Japanese Symposium: Remote Sensing in Marine Biology, Fishery Resources, Oceanography*, (eds M. Petit and J.M. Stretta) Bulletin Institute Oceanographique, Monaco, Vol. 6, pp. 199–207.

Tsai, C.-F. and Chai, A.-L. (1992) Short-term forecasting of the striped bass (*Morone saxatilis*) commercial harvest in the Maryland portion of Chesapeake Bay. *Fisheries Research*, **15**, 67–82.

Tsai, C.-F., Wiley, M. and Chai, A.-L. (1991) Rise and fall of the Potomac River striped bass stock: A hypothesis of the role of sewage. *Transactions of the American Fisheries Society*, **120**, 1–22.

Warwick, C.J., Mumford, J.M. and Norton, G.A. (1993) Environmental management expert systems. *Journal of Environmental Management*, **39**, 251–70.

Waterman, D.A. (1986) *A Guide to Expert Systems*, Addison-Wesley, Reading, Massachusetts.

White, H. (1989a) Neural network learning and statistics. *AI Expert*, December, 48–52.

White, H. (1989b) Learning in artificial neural networks: A statistical perspective. *Neural Computation*, **1**, 425–469.

Chapter three

Potential for geographical information systems (GIS) in fisheries management

Geoff J. Meaden

3.1 INTRODUCTION

Much of the substance of this chapter is speculative. For reasons which will be made clear, GIS has yet to play any direct role in what might broadly be described as the fisheries industry. Since the time is rapidly approaching when this branch of information technology (IT) will make its presence strongly felt here, it is appropriate both to set out the potential for GIS in fisheries research and development and to suggest ways in which its implementation might begin and progress. The reasons for this confidence in GIS are rooted in the observed benefits which it has brought to the management of many terrestrial-based activities, and to the consequent accelerating expansion in the progress and fortunes of the whole GIS sector.

3.1.1 Explanation of GIS

GIS may be defined as the integration of computer hard- and software with spatially referenced digital data so that storage, retrieval, manipulation, analysis, and presentation of the data is possible in order to produce new spatially related output. The input data may be in any digital, textual, map, or numeric form which is capable of being integrated by either a mono-genous (single system) GIS software, and its associated hardware, or in a heterogenous (multiple system) of perhaps two GISs linked to other software and the requisite hardware. GIS output can be either hardcopy or digital in textual, graphical, tabular, or mapped format. Some definitions

Computers in Fisheries Research.
Edited by Bernard A. Megrey and Erlend Moksness.
Published in 1996 by Chapman & Hall, London. ISBN 0 412 59550 8

also consider liveware (operators) to be an essential part of GIS, although this is difficult to justify, in that any human-devised system would involve an element of manual interaction.

GIS is a branch of IT which has only recently become prominent in public or private applications areas, although its practical evolution has been taking place over the past three decades. GIS grew out of developments in Canada during the early 1960s, when it was realized that computers could produce output whose graphical form represented spatial variations in the values of given parameters. Sophisticated output was available in the early 1970s from packages such as SYMAP, GRID, and GEOMAP (Tomlinson, 1989). The comparatively slow, but now accelerating, evolution of GIS since then has resulted from two main factors:

(a) In order to reproduce output which is both detailed and accurate in continuous 2-D space, a large volume of data is needed. In the recent past a surge of data has become available, mainly from remote sensing (RS) and mapping agency sources, and also from the technological ability to scan in mapped data. Much of this data is now becoming acceptably priced, especially since technological advances in computer hardware have dramatically increased the performance:cost ratio of the necessary computer processing.
(b) GIS success has relied upon parallel developments in a number of fields, i.e. computer aided design, RS and image analysis, spatial analysis, digital cartography, surveying, and geodesy. The merging of these fields within a total IT context has been complex, but recent success has spurred significant additional interest and thus a proliferation of research and development, resulting in positive feedback and a reinforcing of success.

The current size of the global GIS market (hardware and software but excluding data) is in excess of US$2.8 billion (CCTA, 1993), with an annual growth rate of approximately 14%.

In most countries the public sector has played the largest part in GIS progress. This is because governments fund many research institutions and universities, and because government departments generate and use the necessarily large data volumes. Additionally, the role of GIS in spatial management is strongly tied to public demands through the democratic process and to the fact that governments may control spatial domains where conflicting use requires careful management, e.g. forestry, recreation, natural resource development, land use planning, and infrastructure. To put it simply, all governments have a major responsibility to achieve spatial harmony.

This current surge in GIS interest has not been made without some major problems. Chief among these have been the legal considerations of copyright and data ownership, problems associated with the structure or format of data inputs, i.e. the need for a standard data transfer format, the high cost of much data, organizational and institutional acceptance of GIS, the verification of the utility of GIS, and lack of suitably trained GIS personnel.

3.1.2 GIS functioning and systems requirements

It is useful to view GIS in terms of inputs, processes, and outputs. Figure 3.1 shows the overall systems functions in its simplest form. The processes within the GIS box, i.e. the capturing of the data and the functions performed by a GIS, are described in this section. Primary and secondary data inputs and acquisition are reviewed in Section 3.3, and the actual and potential information outputs in Sections 3.4 and 3.5.

Data may be entered (captured) from the keyboard, digitizers, scanners, CD-ROMs, computer-compatible tapes, floppy disks, or via networking from external sources. To be of value to the GIS, the captured data must be classified and structured, and then held in a database in such a way as to be meaningful to both users and to the particular GIS software being used. GISs typically have two kinds of databases, i.e. graphical (containing the data relating to mapped coordinates) and nongraphical (containing attribute, labels, descriptive, or features coding). The one imperative for data used in a GIS is that it must be georeferenced. This georeferencing may be in terms of topographic grid references, latitudes and longitudes (for exact locations), or post or Zip codes, street or other area names, census divisions, etc. (for generalized areal units).

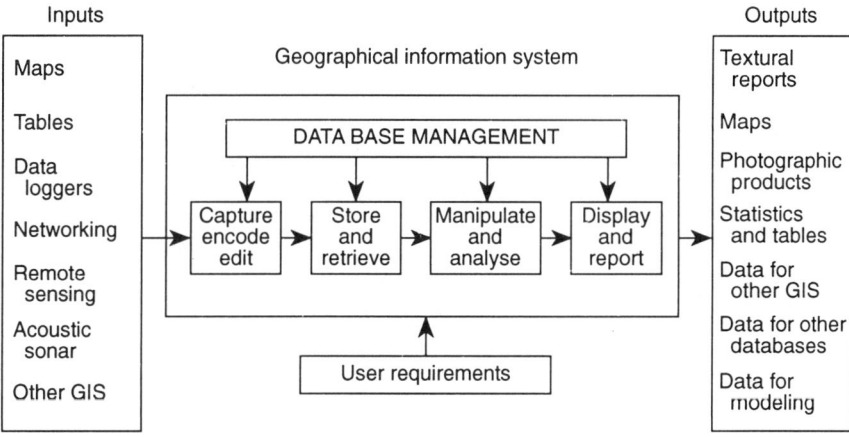

Fig. 3.1 Systems diagram illustrating basic GIS functionality.

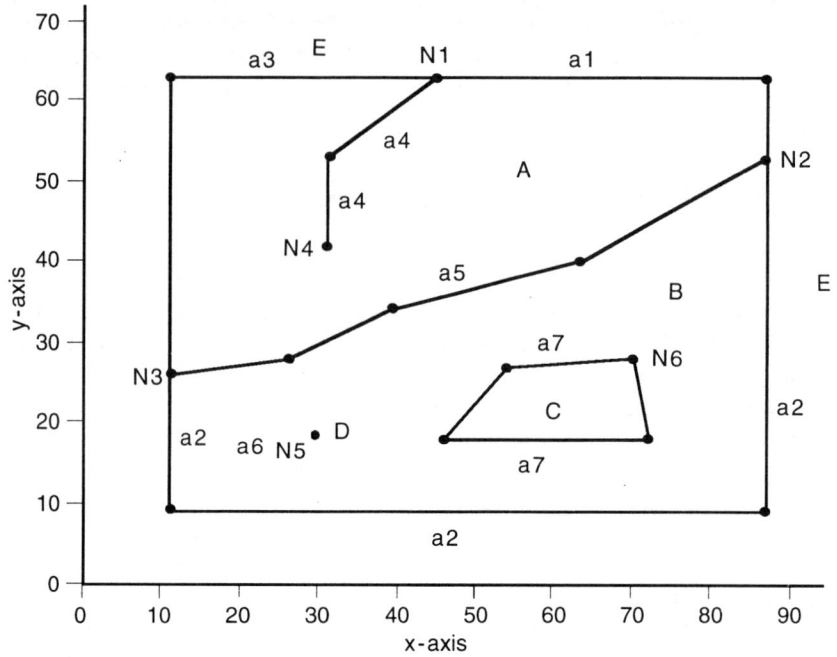

Topological encoding

Polygon topology	
Polygon	Arcs
A	a1, a5, a3
B	a2, a5, 0, a6, 0, a7
C	a7
D	a6
E	area outside map

Arc topology				
Arc	Start node	End node	Left polygon	Right polygon
a1	N1	N2	E	A
a2	N2	N3	E	B
a3	N3	N1	E	A
a4	N4	N1	A	A
a5	N3	N2	A	B
a6	N5	N5	B	B
a7	N6	N6	B	C

Node topology	
Node	Arcs
N1	a1, a3, a4
N2	a1, a2, a5
N3	a2, a3, a5
N4	a4
N5	a6
N6	a7

Arc coordinate data			
Arc	Start x, y	Intermediate x, y	End x, y
a1	45, 63	87, 63	87, 53
a2	87, 53	87, 9; 11, 9	11, 26
a3	11, 26	11, 63	45, 63
a4	45, 63	31, 53	31, 42
a5	11, 26	27, 28; 40, 35; 63, 40	87, 53
a6	30, 18		30, 18
a7	70, 28	73, 17; 46, 17; 54, 27	70, 28

For use in a GIS digital data must be structured in one of two formats – vector or raster. In the vector format all mapped data can be visualized as being either lines, points, or polygons. Data is stored as strings of x,y coordinates which follow lines on a map, or as a single point coordinate representing an object such as a city, village, or lighthouse, depending on the map scale (Figure 3.2). In the raster format the whole mapped surface is composed of a grid of cells whose size determines the final resolution of the map (Figure 3.3). Each cell is allocated a value (that may be related to a color coding) which accords with the feature on the original map. Both data structures have their own advantages, as shown in Table 3.1. Increasingly, GIS packages are able to handle both data structures, although with their rapidly enhanced capacity to handle vast data inputs and to output mapping at a high resolution, GISs are likely to favor raster structuring in the future.

Capturing data in vector format requires not only that coordinates are recorded, but also that attributes (labels or features) are stored and that a topological reference identifier is given so that the computer can recognize the spatial relationship between the attribute and the vectorized lines (Figure 3.2). In raster format feature identification is easy since cells are simply allocated numbers which act as codes for different labels. Because both raster and vector digital maps may use large volumes of data, various storage structures within most GIS data-capture software can allow for efficient compression of the data (Figure 3.3).

GIS software programs are capable of performing a huge variety of functions. The list on page 47 gives a selection of the important functions, although each software package will have its own range. (Specialist users of GIS normally work in cooperation with a software supplier to produce a set of functions commensurate with user requirements.) Each functional operation is designed to produce recalculated data by using an algorithm specifically constructed to yield the desired mapping output. The data itself needs constant managing and maintenance; as Figure 3.1 shows, this is enacted with the use of a database management system (DBMS). These are frequently incorporated within the GIS software package, although a separate relational or object-oriented DBMS may also be required. During data manipulations, output will usually be to the visual display unit (VDU), but final output can also be directed to any external storage or hard-copy capture device, i.e. the latter ranging from desk-top printers (dot-matrix,

---

Fig. 3.2 An example of the vector format illustrating point (node), line (arc) and polygons. The vector polygon, arc and node topological encodings of this map are also shown. The arc coordinate table links the positions of all these features to their real-world locations.

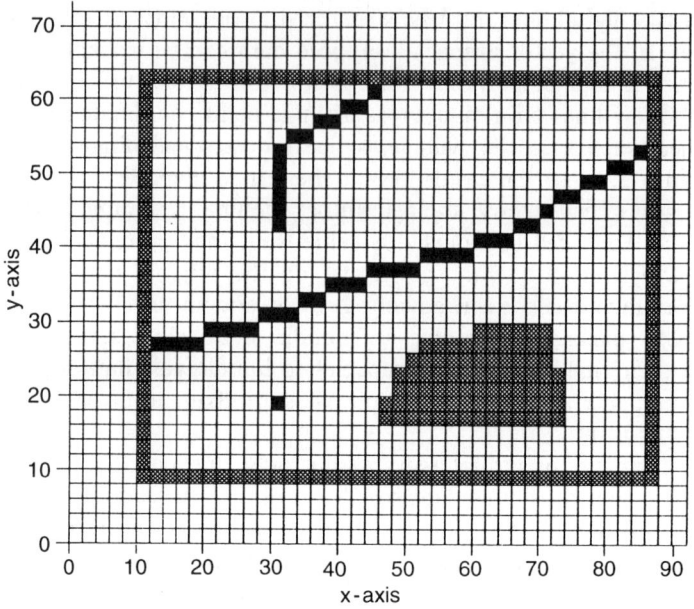

Raster run-length code structure

Row	Run-length encoding
66	0, 90, 0
64	0, 90, 0
62	0, 8, 0; 10, 86, 1; 88, 90, 0
60	0, 8, 0; 10, 1; 12, 42, 0; 44, 2; 46, 84, 0; 86, 1; 88, 90, 0
58	0, 8, 0; 10, 1; 12, 38, 0; 40, 42, 2; 44, 84, 0; 86, 1; 88, 90, 0
56	0, 8, 0; 10, 1; 12, 34, 0; 36, 38, 2; 40, 84, 0; 86, 1; 88, 90, 0
54	0, 8, 0; 10, 1; 12, 30, 0; 32, 34, 2; 36, 84. 0; 86, 1; 88, 90, 0
52	0, 8, 0; 10, 1; 12, 28, 0; 30, 2; 32, 82, 0; 84, 2; 86, 1; 88, 90, 0
50	0, 8, 0; 10, 1; 12, 28, 0; 30, 2; 32, 78, 0; 80, 82, 2; 84, 0; 86, 1; 88, 90, 0
48	0, 8, 0; 10, 1; 12, 28, 0; 30, 2; 32, 74, 0; 76, 78, 2; 80, 84, 0; 86, 1; 88, 90, 0
46	0, 8, 0; 10, 1; 12, 28, 0; 30, 2; 32, 70, 0; 72, 74, 2; 76, 84, 0; 86, 1; 88, 90, 0
44	0, 8, 0; 10, 1; 12, 28, 0; 30, 2; 32, 68, 0; 70, 2; 72, 84, 0; 86, 1; 88, 90, 0
42	0, 8, 0; 10, 1; 12, 28, 0; 30, 2; 32, 64, 0; 66, 68, 2; 70, 84, 0; 86, 1; 88, 90, 0
40	0, 8, 0; 10, 1; 12, 58, 0; 60, 64, 2; 66, 84, 0; 86, 1; 88, 90, 0
38	0, 8, 0; 10, 1; 12, 50, 0; 52, 58, 2; 60, 84, 0; 86, 1; 88, 90, 0
36	0, 8, 0; 10, 1; 12, 42, 0; 44, 50, 2; 52, 84, 0; 86, 1; 88, 90, 0
34	0, 8, 0; 10, 1; 12, 36, 0; 38, 42, 2; 44, 84, 0; 86, 1; 88, 90, 0
32	0, 8, 0; 10, 1; 12, 32, 0; 34, 36, 2; 38, 84, 0; 86, 1; 88, 90, 0
30	0, 8, 0; 10, 1; 12, 26, 0; 28, 32, 2; 34, 84, 0; 86, 1; 88, 90, 0
28	0, 8, 0; 10, 1; 12, 18, 0; 20, 26, 2; 28, 58, 0; 60, 70, 1; 72, 84, 0; 86, 1; 88, 90, 0
26	0, 8, 0; 10, 1; 12, 18, 2; 20, 50, 0; 52, 70, 1; 72, 84, 0; 86, 1; 88, 90, 0
24	0, 8, 0; 10, 1; 12, 48, 0; 50, 70, 1; 72, 84, 0; 86, 1; 88, 90, 0
22	0, 8, 0; 10, 1; 12, 46, 0; 48, 72, 1; 74, 84, 0; 86, 1; 88, 90, 0
20	0, 8, 0; 10, 1; 12, 46, 0; 48, 72, 1; 74, 84, 0; 86, 1; 88, 90, 0
18	0, 8, 0; 10, 1; 12, 28, 0; 30, 2; 32, 44, 0; 46, 72, 1; 74, 84, 0; 86, 1; 88, 90, 0
16	0, 8, 0; 10, 1; 12, 44, 0; 46, 72, 1; 74, 84, 0; 86, 1; 88, 90, 0
14	0, 8, 0; 10, 1; 12, 84, 0; 86, 1; 88, 90, 0
12	0, 8, 0; 10, 1; 12, 84, 0; 86, 1; 88, 90, 0
10	0, 8, 0; 10, 1; 12, 84, 0; 86, 1; 88, 90, 0
8	0, 8, 0; 10, 86, 1; 88, 90, 0
6	0, 90, 0
4	0, 90, 0
2	0, 90, 0

Table 3.1 Comparison of raster and vector data structures

GIS Feature	Raster	Vector
Data capture	fast	slow
Data volumes	large	small
Graphic display	poor	good
Data structure	simple	complex
Geometrical accuracy	low	high
Linear network analysis	poor	good
Area/polygon analysis	good	poor
Combining data layers	good	poor
Map generalization	simple	complex
Topology	complex	easy

laser, or ink-jet) to a variety of specialist mechanical, thermal, or electro-static plotters. Output is usually on paper, but it may be to film or other media.

(a) Data input and encoding:
   (i)   data capture, e.g. digitizing and integration of external data;
   (ii) data validation and editing, e.g. checking and correction;
   (iii) data structuring and storage, e.g. constructing surfaces and data coding.
(b) Data manipulation:
   (i)   structure conversion, e.g. conversion from vector to raster;
   (ii) geometric conversion, e.g. map registration, scale changes, projection changes, map transformations, rotation;
   (iii) generalization and classification, e.g. reclassifying data, aggregation or disaggregation, coordinate thinning;
   (iv) integration, e.g. overlaying or combining map layers;
   (v) map enhancement, e.g. image edge enhancement, add title, scale, key, map symbolism, draping overlays;
   (vi) abstraction, e.g. calculating area centroids, Thiessen polygons;
   (vii) buffer generation, e.g. calculating and defining corridors.
(c) Data searching and retrieval:
   (i)   selective retrieval, e.g. based on user-defined themes, or browsing, querying, windowing;

Fig. 3.3 The raster data format for the same mapped entities as shown in Figure 3.2. The simple data compression technique of run-length encoding is used to save on data storage space.

    (ii)   spatial searching, e.g. on points, lines and areas to determine distances, angles or overlaps.

(d) Data analysis:
    (i)   spatial analysis, e.g. connectivity, proximity, contiguity, intervisibility, interpolation, digital terrain modeling;
    (ii)   statistical analysis, e.g. histograms, correlation, measures of dispersion, frequency analysis;
    (iii) measurement, e.g. line length, area and volume calculations, distance and direction.

(e) Data display:
    (i)   graphical display, e.g. maps, graphs;
    (ii)   textual display, e.g. reports, tables.

(f) Database management
    (i)   support and monitoring of multiuser access to the database;
    (ii)   coping with systems failure;
    (iii) communication linkages with other systems;
    (iv) editing and updating of databases;
    (v)   organizing the database for efficient storage and retrieval;
    (vi) maintenance of database security and integrity;
    (vii) provision of a 'data independent' view of the database.

As with other branches of computing, GIS programs and functionality have rapidly adapted to changing computer processing capabilities. So, whereas until the late 1980s most software was designed to perform on minicomputers or mainframes, nearly all GIS packages now have microcomputer or workstation versions. This means that it is now possible to operate a complete GIS (hardware and software) for little more than US$1500, e.g. by using a minimal configuration consisting of perhaps an 80386 processor with 4 MB of RAM and a small hard-drive, plus a desk-top monochrome printer and a raster-based software package such as IDRISI or OSU-MAP. It would be fair to say that, because of their affordable, high level processing capacity, UNIX-based workstations would be the major commercial GIS vehicle now being used. Although these stand-alone systems can be assembled using an almost infinite hardware configurations mix, more sophisticated GISs usually operate as local area networks (LANs), or less frequently as wide area networks (WANs).

3.2 VALUE OF SPATIAL ANALYSES

Before discussing the use of GIS in marine fisheries, it is relevant to pose the question 'what is the value of the spatial perspective and of spatial analyses, both generally and to marine fisheries in particular?' In this section the discussion is limited to the broader, general concepts; Section 3.5 outlines in more detail the potential that GIS offers to marine fisheries. The general value of the spatial perspective is best approached from the three viewpoints which now follow.

3.2.1 Facets of spatial variation

Since the Earth is not composed of ubiquitous physical phenomena, there is complex variation in the ordering of natural space. This idea can be interpreted not only in terms of fixed physical features but also of moving organic and inorganic phenomena. There are variations in the disposition of mountain ranges, flood plains, tropical rain forests, ocean currents, whales, etc. This diversity has given rise to variations in human spatial patterns of occupation and activity. The concept of spatial variation itself can be extended to encompass not only the actual disposition of phenomena, i.e. where things are located, but also to a variety of other considerations and dimensions which incorporate realities such as time, movement, change, and inequity. Given these various facets of spatial variation, it is both interesting and relevant to postulate a simple taxonomy of spatial variation (Table 3.2).

Clearly, the facets overlap; Table 3.2 makes the case that the viewing of spatially related issues is not necessarily straightforward, e.g. within a given geographic area there may be various spatial facets which all require consideration. To take a marine fisheries example, there could be an area of water in which several trawlers were operating, each targeting different species, and perhaps each having varying quotas and rights of access to each species. Clearly very different considerations need to be given when

Table 3.2 Taxonomy of spatial variation

Facet of spatial variation	*Examples*
Disposition or distributions	dispersal of a species, location of ports
Diversity	sea-bottom types, species variety in an area
Changes in distributions	temporal change in species mix, changing pattern of stocks
Human movement patterns	fishing vessel routing, in-harbour movements
Rights of access	restricted fishing zones, exclusive economic zones
Inequities	densities of plankton blooms, ability to command resources
Natural movement patterns	ocean currents and winds, fish migration routes
Management strategies	degree of future planning, purposeful interventions
Conflicts of interest	waste disposal versus fishing, sewage outfalls versus coast zone

assessing the criteria for mapping, or for spatial allocation, of these various facets. Some facets may undergo rapid changes, others remain perfectly static, some require legal backing, and others require periodic changes. So there exists an extraordinary variety of ways in which spatially related issues are fundamental to marine fisheries and therefore to their management.

3.2.2 Optimization of location

It has long been recognized that a main factor controlling the success of many activities is location optimization. Basically, this can be thought of in terms of organizing space so that it functions (or is organized) in a way that best approaches an ideal. The notion of 'ideal' is typically envisaged in an economic sense, although it is increasingly being thought of in terms of environmental or social criteria. From an economic viewpoint Alonso (1960) conceived of urban economic space as being organized in terms of activities which allowed some areas to attract higher rents than others (Figure 3.4). Similarly, Von Thunen (1826) and Weber (1909) drew up models which showed, respectively, how agricultural and industrial space might also be organized on the basis of changing activity use with distance from a central location, i.e. as dictated by increased transport costs.

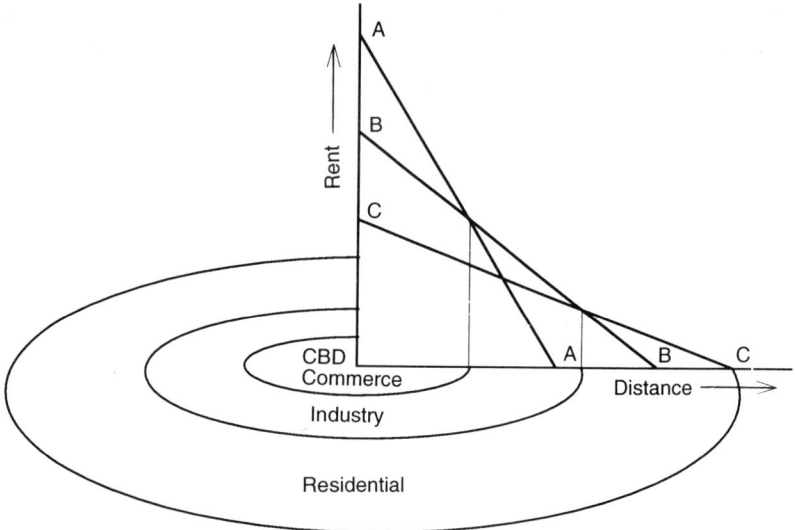

Fig. 3.4 The bid rent curve as conceived by Alonso (1960). Higher rents in the center are based upon the central business district's (CBD) greater accessibility: A–A, rent that commerce will pay; B–B rent that industry will pay; C–C rent that residential will pay.

The rise in social and environmental awareness has required the recognition of the fundamental importance of organizing space, largely through local and regional planning processes, so that an agreed balance is obtained in meeting what are often conflicting spatial demands. From Table 3.2 it is easy to see the potential conflicts over marine space. This means that allocation of activities to areas, in order to achieve some level of spatial optimization, will be increasingly necessary if conflicts are to be minimized and if maximum success is going to be achieved for most marine activities. Spatial allocation is a type of demand for which a GIS is particularly suited.

3.2.3 Efficiency of graphical perception

There is a widespread philosophical agreement, which certainly accords with common sense, that the spatial aspects of all existence are fundamental ... space seems to be that aspect of existence to which most other things can be analogized or with which they can be equated. (Robinson and Petchenik, 1977)

In addition to the philosophical viewpoint, it is clear that the human brain is better able to perceive 2- or 3-D graphical images than, for instance, pure linguistic or numeric images. Thus Figure 3.5 shows a simple gradient which places facets of perception on a hierarchy from reality to abstraction. Maps occupy a huge range on the hierarchy. This is because they can vary from a realistic image, such as a highlighted aerial photograph, via a range of topographic and thematic maps to quite simplistic or abstract forms, e.g. the topological connectivity maps of a railway system.

Though not as easy to perceive as photographs or cine-photography, maps have distinct advantages over written textual information. Reading text requires a logical progression through the whole script in order to get the essence of what is being communicated, whereas the map reader can selectively extract information in more or less any order, and in varying degrees of detail. Maps allow for the visualization of relative location or relationships which correspond exactly to real space. In doing this they allow easy recovery of distance, direction and aerial measurements. A huge range of symbolism can be applied to point, line, or spatially extensive features, and maps can be selective in what they choose to display. Arising from the seminal work of Kolacny (1968) and his successors, there is abundant evidence that facts relating to spatial distributions are better envisaged and understood if relevant maps are available.

Since the static nature of a hard-copy map is no longer a barrier, mapping by means of the computer and VDU greatly enhances a map's utility. Interaction is possible with the displayed image such that not

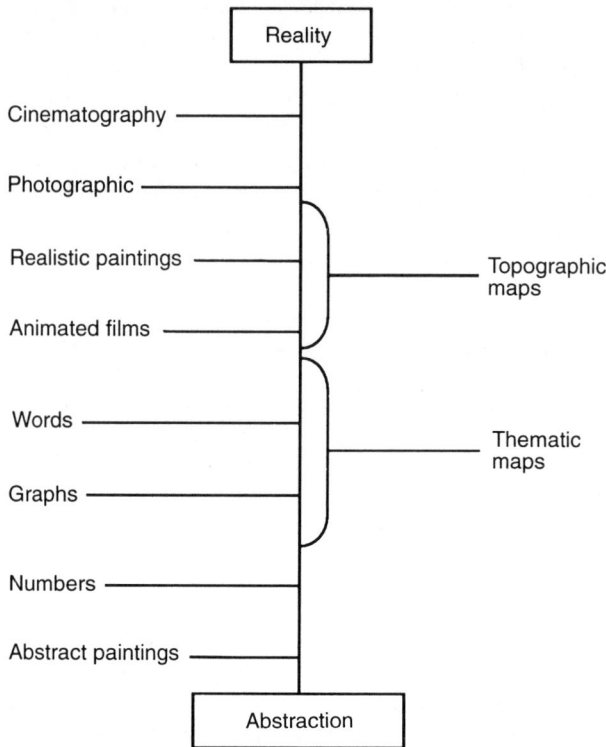

Fig. 3.5 A simplified hierarchy of human perceptions. Nonmapping entities are shown in their relative positions, though each may occupy a limited spectrum of the hierarchy. The two main mapping types are shown as occupying their approximate range of perceptions.

only can changes, edits, updates, etc., be instantly accomplished, but maps can also be reoriented, scale changes can be made via zoom facilities, and temporal sequential modeling can be applied. The advent of multimedia systems, integrated with GIS, will soon give the potential to present complex mapped information in a range of perceptual modes, e.g. maps will become 'alive' incorporating sound, added visual images, or perhaps animations to show sequential temporal or value changes. Spatial visualization via computers is offering the human mind the ability to almost instantly conceptualize the organization of any of the taxonomic facets shown in Table 3.2 in an unlimited number of ways. Whether the mind can cope with the resulting information overload is another matter!

3.3 DATA SOURCES FOR A FISHERIES GIS

It must be clear by now that, for efficient and accurate functionality, any GIS will require a vast data input. In fact, with the rapidly falling prices of computing hard- and software, data acquisition is now the major consideration with regard to GIS implementation. Not only must data costs be considered, but also the types of data required, data sources and locations, methods of data collection (including sampling strategies and the equipment required), data formats and structures, data copyright, and the reliability of the data. This section highlights some aspects of data acquisition which are of particular significance. Some of the appropriate methods for inputing this data were considered in Section 3.1.2.

Before data is collected, clear aims for a GIS must have been established. Given the plight of the world's fisheries, in many instances these aims will be a response to an urgent need to rectify a deteriorating situation. However, since not all aspects pertaining to a fisheries GIS could be instigated easily or immediately, it is likely that more realistic aims will arise from the fact that some data is easier to obtain than others. Having established project aims, the drawing up of data requirements is usually relatively simple, although the actual derivation of this data can frequently be most frustrating!

Since a GIS requires so much data, obtained from variety of sources, it will be important that a meta database is established in order to monitor both the data holdings and the quality of the data.

Whether use is made of primary (original) or secondary data largely depends upon the nature of the GIS requirements, data availability, and the level of resource funding. In an applications field such as fisheries, with its obviously mobile resource base and environment, much of the data will be of a primary nature, if only in the sense that it will be updating a previously captured spatial scenario. It is likely that most secondary data will only be available for those areas of fisheries concerned with permanent static elements such as shoreline features or in some cases bathymetry.

3.3.1 Acquiring primary data

Primary data may be acquired either manually or automatically. Manual data inputs come from preprinted form surveys, interviews, or questionnaire results which are entered into GIS-compatible spreadsheet or database programs, one of whose fields must be a georeference. At the lower end of the automation scale, GIS data can still be acquired using nonelectronic equipment such as hand-held cameras, site surveying equipment, meteorological instruments, and trawl surveying techniques. The amount of data collected by these means would be limited in the sense that collection is necessarily slow, but in site- or location-specific situations these methods can still be vital. At a higher level of sophistication, much data collection equipment now incorporates microelectronics. This has improved collection by increasing accuracy, portability, speed, and ease of use, while frequently reducing price. The simplest equipment includes a range of devices which provide direct readouts on temperatures, weights, flow rates, water qualitative parameters, etc., but more sophisticated devices are now able to store information for later downloading, usually for export to another computer software package (Price, 1992). These data loggers typically incorporate software which has been designed for recording a limited number of parameters, e.g. hand-held laser survey equipment which records azimuth range, height, inclinations and coordinates, or fixed location instruments which monitor meteorological or water quality variables. Data obtained can be numeric values or attribute information, and intelligent loggers can switch between sampling strategies or perform certain mathematical transformations. The variable functions which programmable data loggers now perform are innumerable and, as a method of data acquisition, they look certain to revolutionize GIS.

Another item of equipment which is having a great impact on GIS is the global positioning system (GPS). This is a hand-held or vehicle-mounted device which, through transmission linkages to specific US NAVSTAR satellites, is able to establish the operator's latitude and longitude position on the planet to submeter accuracy if used in its differential mode (Gilbert, 1994). Since an experienced operator can now compile several thousand GPS readings a day, the potential for location recording and subsequent marine mapping is huge. Fishing or survey vessels can accurately establish where trawl hauls were set and recovered, obstacles can be mapped, water quality sampling points can be read, and position can be established relative to legal or other boundaries. GPS will allow for the storage and export of numeric, attribute, and positional data direct to most GIS packages. There are already commercial marine equipment companies who are offering GPS base receiver facilities for fishing vessels in western European waters, and many countries are making GPS equipment obligatory on fishing vessels of over a certain size.

At a still higher level of sophistication, a piece of electronic equipment which has long been used for fisheries purposes, the acoustic SONAR system, is now being linked to a large number of computer packages including GIS. From a GIS viewpoint it is useful to know the density and distribution patterns of fish within a survey area, though echo sounding can also be used to determine bathymetry, and acoustic devices now entering the market will allow for the mapping of bottom sediment types. A final sophisticated primary data collection technique is via the use of aerial (photography or videography) or satellite RS methods. These cannot be described here, but Allan (1992) gives a useful summary covering marine applications of RS, and Maussel *et al.* (1992) give details on airborne videography.

3.3.2 Acquiring secondary data

Our examination of primary data collection followed a progression based upon the equipment needed for this data acquisition. With regard to acquiring secondary data, considerations of equipment are of less concern. Instead, the importance lies in the sources of data, and to a lesser extent its form and format. Regarding the latter it is simply necessary to state that the secondary data suitable for a GIS will consist mostly of maps, textual information and tabular or graphical data. It is likely to be held as hard copy, on film, or sometimes now in digital format. With the advent of on-line search facilities, searching for secondary data is theoretically becoming easier, although with the costs involved, plus the proliferation of literature which is often in the 'grey area', this is not always the case.

For all GIS, maps clearly form a fundamental data source. From the fisheries perspective it is logical to classify existing maps under three headings:

(a) *Hydrographic maps* – The maritime authorities or government mapping agencies of most countries produce hydrographic maps which should meet certain standards as set by the International Hydrographic Organisation. These maps vary in scale from 1:50 000 to 1:10 000 000, although more detailed maps exist for special areas. They concentrate on coastal zones, to show not only shoreline features but buoys, obstacles, bathymetry, and other navigational information. The Hydrographic Office in the UK provides authoritative hydrographic charts for most coastal areas of the world; these are now all available in digital format.

(b) *Topographic maps* – For land-based information, national mapping agencies of most countries provide topographic maps in scales which vary from about 1:1000 to 1:1 000 000. For most GIS requirements these are reliable data sources, although the infrequency of update and lack of detail must be borne in mind, especially for more remote or less-developed areas. In many regions, aerial or satellite remote sensing is rapidly being applied to map updating, and in the more developed countries most map series have been partly or fully digitized.

(c) *Thematic maps* – These are specialized maps which usually concentrate upon one particular theme. Thematic maps form the largest growth area in mapping, not only because of the increased recognition of their use-fulness, but also because computer graphics has allowed for their pro-liferation and because of the growth in specialized publications which

utilize mapping as an illustrative form. Still another reason is the increased emphasis now being placed on spatial planning.

Once hard-copy maps have been acquired they must be converted to a
digital format. Digitizing has traditionally been carried out by the final GIS
user, who can readily define the exact outlines needed. Increasingly,
however, pre-digitized data may be bought from specialist agencies, e.g.
topographic outlines from government mapping agencies, or scanning is
favored since scanner prices have dropped compared with the high costs
associated with manual digitizing. Digitized data is usually captured in
layers, with each representing one feature type. Figure 3.6 shows a
sequence of data layers which may be appropriate to a coastal zone GIS.

Tabular data sources usually consist of a variety of aggregated information, from single tabular sheets to huge volumes of collated forms. Such
information is increasingly held on computer databases. Integrating tabular

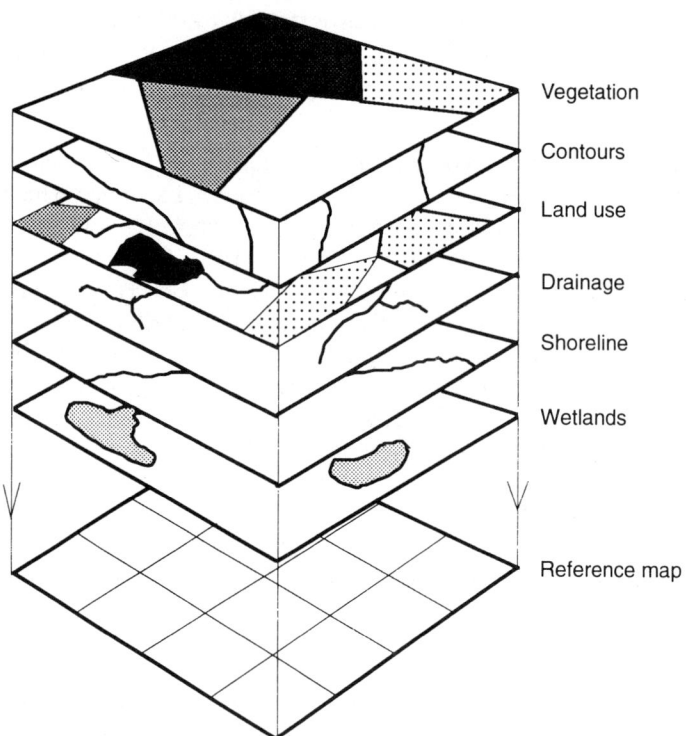

Fig. 3.6 Layers used to construct an integrated coastal resource management
database for part of Georgia, USA. In built-up areas it might be necessary to add
layers which reflected greater human impact (modified from Welch *et al.*, 1992).

data into a GIS is usually far from satisfactory, unless a data gathering system has been specially implemented with GIS in mind. Data age, level of aggregation, lack of a georeference, confidentiality, dubious levels of accuracy, and difficulties in gaining access can all pose problems. It is therefore usually advisable to set up purposefully planned tabular databases.

Networking provides an increasingly efficient way of gaining access to existing digital databases. Until recently, although a huge number of databases existed, there was no easy way of finding out what they consisted of, where they were held or how to access them, (e.g. via exchange or disk or tape, or via networking). Specialist data information networks are now appearing via the electronic Internet, and gateways such as Gopher and the World Wide Web. The use of the network requires an electronic mail (E-mail) address, a computer with an appropriate communication interface, plus a modem link to other computers through existing telephone or other dedicated data lines. The user can then obtain access to a huge range of data via bulletin boards or specialist subscriber lists such as Coast-GIS, Fish Ecology or GIS-List. Although these are presently based in, and their data relates to, developed countries, wide area networking will eventually provide a relatively simple medium for world-wide data diffusion. However, there are still barriers to be overcome before networking becomes truly international, e.g. standards for data exchange systems, better data compression ratios, the installation of fibre-optic transmission lines and the implementation of the Integrated Services Digital Network, which will better interactively link all telephone systems. A recent publication (Bostock, 1994) is the first to specifically cover marine and aquaculture networking possibilities. The geographic area most advanced in marine-based networking capabilities is the eastern seaboard of Canada (Butler and LeBlanc, 1990).

Using the same principle as networking, many ships are now equipped with electronic chart display and information systems (ECDIS) which can not only display the ship's position and any navigational information, but can also import any GIS-based data, GPS data, and updated navigation data via the INMARSAT satellite communication linkages (Bernhardsen, 1992). Now that methods have been found for the compression of huge amounts of data, the transmission of the final GIS maps will soon be possible via networking (Knott and Shiers, 1994).

Some digital data is available via means other than networking, e.g. some CD-ROMS provide outline maps or tabular data, and government departments are sometimes able to supply relevant data on disk. Satellite RS data of interest to fisheries is available from a number of agencies. It includes data taken from both polar orbiting and geostationary satellites collected since the early 1970s. There are problems of high costs, cloud-obscured imagery, poor resolution, the necessity for ground truthing, etc. However, with satellite activity poised to grow during the next decade, and with much special attention being given to gathering marine-based data (such as that from ERS-1), comparative costs are likely to fall for fisheries-relevant data. Additionally, new image data searching facilities, operated by the European Space Agency in conjunction with the Centre for Earth Observation, are about to come on line which will provide a 'queryable' access entry point to multiple inventories of satellite imagery, allowing this imagery to be better located and defined and therefore to gain in its utility.

3.4 PRESENT USES OF MARINE GIS

Section 3.1.1 gave a hint that GIS has been widely adopted elsewhere as a potent management tool in both the private and public sectors. GIS is now being extensively adopted in marine-associated activities. However, there are no true marine fisheries GISs, i.e. systems whose prime function is to record and manage fisheries, although the potential value is now being recognized (Simpson, 1992; Li and Saxena, 1993; Meaden, 1993). Some attempts have been made to utilize GIS in inland fisheries and aquaculture; examples are given in Meaden and Kapetsky (1991). Here we first examine why the adoption process has been so slow and then briefly illustrate some of the ways in which GIS is slowly permeating marine or fisheries-related usage.

3.4.1 Why no marine fisheries GIS?

Given the advantages that GIS has to offer, its rapid adoption in other activities, and the dire plight which many of the world's fisheries face, and given the fact that marine fishing is the world's most spatially extensive activity, then it may seem extraordinary that GIS has not been taken up with enthusiasm by at least those whose task it is to manage fisheries. The reasons for this apparent lack of enthusiasm are best viewed separately both from the private and the public perspective.

In the private sector the fishing industry is organized in an extremely fragmented way. Not only are there a multitude of different sectors, but the actual fishing activity functions at scales varying from perhaps one person operating a small part-time unit up to large companies operating fleets of international trawlers. Fragmentation is reinforced by the fact that marine fishing must be marginally located around what may be a very long coastline, thus greatly reducing the chance for agglomerative linkages or other forms of cooperation. This lack of cooperation is reinforced by the fact that freedom of access to the resource causes intense competition. Also, since fishermen seldom have rights to particular spatial areas, there is no incentive for the individual to manage an area for longer-term benefits. So clearly, from the viewpoint of the private fisherman or fishing company, a fisheries GIS would have little functionality.

The outlook is different in the public sector. A fisheries GIS is more likely to operate here since there could be, and frequently is, more centralized control and the incentives to manage the fishing industry are very high indeed. Also, in the public sector there is frequently access to a wide array of data from variable sources, and governments have powers to legislate for additional data collection. But despite these incentives, most governments are a long way from managing their fisheries with the aid of GIS. Why is this?

In developing countries, reasons relating to costs, data, and priorities are paramount. Although the actual implementation and management of a fisheries GIS *per se* might not be expensive, the almost total lack of appropriate data can only be remedied by either expensive purchase from outside sources or contractors, or setting up expensive (and usually extensive) data gathering systems. Most fisheries departments in developing countries, whether incorrectly or realistically, would perceive a fisheries GIS as either a low priority or at best a doubtful luxury.

Lack of GIS adoption by fishery departments in the developed world has more subtle and varied reasons. First, here too there would often be financial limitations, especially if existing data collection systems were at a crude level. Secondly, fishery departments are frequently very fragmented, being divided into numerous small, almost independent, departments which carry out a diverse range of activities, many of which are not specifically spatially related. Thirdly, there is seldom anyone in a position to act as an innovator, who could implement and nurture a GIS installation. This being the case, it is difficult for a comparatively new IT system to prove its efficacy. Fourthly, even if data gathering systems were in place, it is likely that past data will not have been gathered in an appropriate format and the efforts required to centralize, structure, and standardize this data would be enormous. Much data may be so generally georeferenced that it would be of only marginal value. Finally, there are undoubtedly both perceived and real difficulties of collecting, managing, modeling and manipulating data which may be referenced to a 4-D and mobile environment.

3.4.2 Ways in which fisheries related GISs are being used

Despite the obstacles listed above, GIS is beginning to permeate into areas which are related to fisheries. Here some of these are outlined, to give a feel for both the complexities and the opportunities offered. Section 3.5 outlines additional functions or database areas in which GIS could be utilized.

Nautical charts

As mentioned in Section 3.3.2, most maritime countries produce nautical charts. Although these charts have been traditionally used by defense departments, commercial shipping, and fisheries, there are now large growth areas in the leisure industry and for various oceanographic purposes. In order to improve levels of accuracy and to cope with the large volumes of acoustically derived digital bathymetric data, hydrographic offices in most leading maritime nations have fully digitized marine chart production. GIS functionality can usefully be employed with this data to create 3-D images of the sea floor, to draw profiles showing relationships between water qualitative parameters and depth, and to allow for the output of charts at user-defined scales for specified areas (Wardle, 1992). Obviously, mapping output need no longer be confined to hardcopy, e.g. fisheries vessels equipped with GIS software such as Intergraph's I/SEA or I/SEAVUE, can utilize an electronic chart display to visualize sounding data as either contour or 3-D representation.

Coastal zone management

The important role of the coastal zone to fisheries, in terms of providing for feeding and sometimes reproductive needs, is increasingly being recognized. At the same time the coastline is under growing pressure, as a greater proportion of the population wish to live and take their recreation in this zone. Being at the interface between land and sea invariably means that the ecology here is complex and is likely to be sensitive to a range of pressures. Given these facts, there has been a huge interest in GIS applications aimed at solving specific management problems. A number of projects have recently been reported, for example:

(a) Welch *et al.* (1992) report on an initiative to evaluate environmental trends over a 40 year period in an area of relatively unaltered salt marsh on the coast of Georgia, USA. They used existing mapping, GPS, photogrammetric aerotriangulation, image analysis, and GIS techniques to build up the sequence of databases on the themes shown in Figure 3.6. They were able to monitor changes in local human activity plus various geomorphological changes, and have importantly shown that their techniques could be applied to other areas needing management.

(b) There are several reports on protecting the Wadden Sea area via the use of GIS, e.g. Liebig (1994) and Schauser *et al.* (1992). Information layers on landscape elements, coastal protection measures, recreation, agriculture, eco-potential, infrastructure, settlement, and legal status have

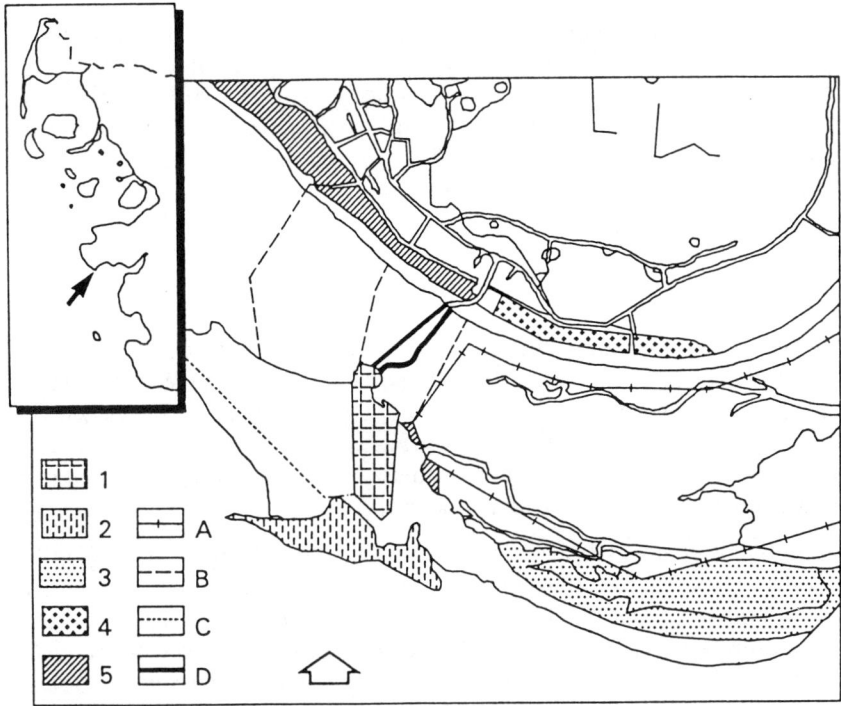

Fig. 3.7 GIS-generated map to show conflicts of recreation and ecopotential near St. Peter-Ording, Schleswig-Holstein National Water Park, Germany: 1, car parking on beach; 2, wheel tracks on beach; 3, bird breeding/trampling; 4, golf course; 5, trampled paths; A, border of the national park; B, specific trampled paths; C, wheel track; D, main access to beach (after Schauser *et al.*, 1992).

been stored. The main intention of the GIS is to identify regions of conflict and of morphological change and to motivate a more integrated and supraregional approach towards physical planning for the coastal region (Figure 3.7).

Apart from these specific applications, there are further reports outlining some of the problems which need to be overcome in applying GIS to coastal zone management. Thus Riddell (1992) underlines the need for an integrated approach to the solution of coastal problems, and he exemplifies how GIS can provide the means whereby a coastal management GIS can be properly structured in a way which is both flexible and practical. Harper and Curtis (1993) report on how the UK Hydrographic Office is working in conjunction with the UK Ordnance Survey on a GIS-based coastal zone mapping project, with the objective of identifying the needs and the market for a new mapping product which incorporates a combination of terrestrial and marine-based data. Although the advantages of using GIS for coastal zone mapping may be fairly obvious, Kam *et al.* (1992) detail the not-inconsiderable problems faced in the developing world. These problems include costs, lack of familiarity with methodologies, lack of technical expertise, and the inaccessibility of data.

Aquaculture location

The early work on the application of GIS to aquaculture was mostly confined to inland aquaculture and fisheries, e.g. Meaden (1987) and Kapetsky *et al.* (1988). This reflected the fact that data was more easily available on the requisite production parameters and that most early commercial aquaculture activities were based on freshwater production. However, there are now available a number of studies which demonstrate the usefulness of GIS to marine aquaculture:

(a) Kapetsky *et al.* (1987) utilized both satellite RS and GIS to identify optimum areas for intertidal mollusc culture, for extensive shrimp and fish culture in salt ponds, and for semi-intensive shrimp culture outside mangrove areas along the Gulf of Nicoya coastline on the Pacific coast of Costa Rica. Like many GIS analyses, his methods relied upon the use of overlay techniques to identify optimum production areas.

(b) Many coastal areas of the Philippines have suffered severe ecologic pressure because of the needs for both fish protein production and forest products. This is manifest in the accelerating rates of mangrove clearance to make way for the shallow brackish ponds for milkfish and shrimp rearing. In a study of a small area of northern Luzon, Paw *et al.* (1992) showed the usefulness of GIS in planning for better management, such that coastal areas could be zoned so as to protect the natural ecology (mangroves and reefs) and such that problems arising from aquaculture in those specific areas could be minimized.

(c) Not all uses of GIS in aquaculture have been for relatively large-scale location purposes. Thus Ross *et al.* (1991) reported the use of an inexpensive raster-based GIS (OSU-MAP) for determining the best location for the siting of salmonid sea cages within a small bay in western Scotland. Here the concern was with factors relating to bathymetry, currents, shelter, wave height, wind dominance, and various facets of water quality. The use of GIS indicated that just 1.26 hectares of the total area were suitable.

Overviews of the application of GIS to aquaculture can be obtained from both Meaden and Kapetsky (1991) and Beveridge and Ross (1991).

Marine water quality

A number of recent studies have utilized GIS to make varying assessments of water-qualitative parameters. Some examples can be briefly described:

(a) Collins and Hurlbut (1993) used a GIS to show the risk of attempting to salvage an oil barge from the bed of the Gulf of St. Lawrence, Canada, which still contained 3100 t of oil. They were able to highlight the seasonal distribution of fish catches and to accurately plot patterns of fishing in relation to suspected oil flows as determined by an oil spill trajectory model (Figure 3.8).

(b) The changing patterns of turbidity in the Zhujiang estuary of southern China were mapped by Lo and Hutchinson (1991) using a number of Landsat and Spot satellite images in conjunction with a raster GIS. The authors concluded that GIS would be a powerful tool in explaining circulatory dynamics in a complex estuarine environment.

(c) Legault (1990) showed how GIS could be used to evaluate the extent to which the closure of areas for shellfish production, due to bacterial contamination, would affect the industry in Prince Edward Island,

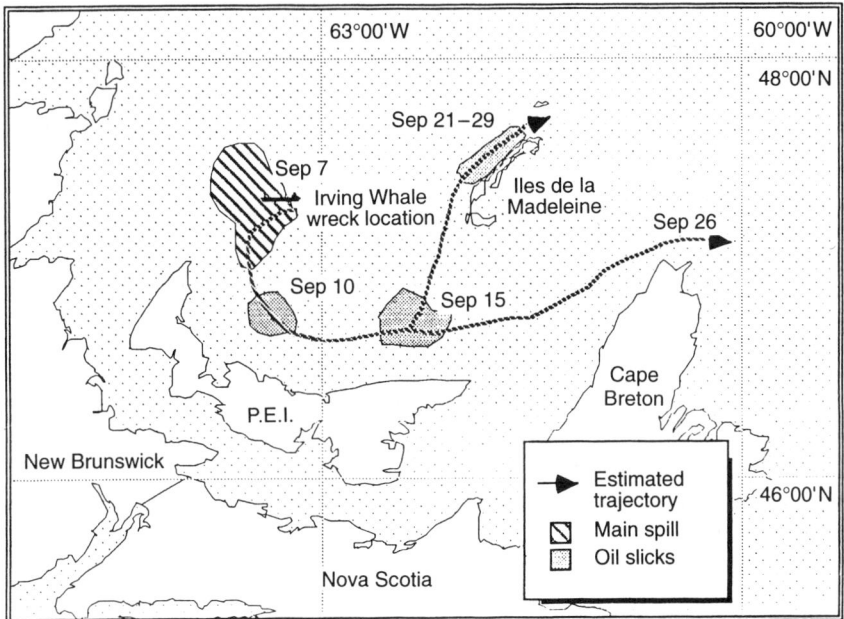

Fig. 3.8 Estimated oil trajectory from the site of the original wreck location. The main spill is the area of sea covered by oil immediately after the September 7 sinking. Positions of this slick are shown for three other time periods (after Collins and Hurlbut, 1993).

Canada. Thus, knowing the prices received, the areas of shellfish growing leases, and where the contaminated zones were located, it was possible to forecast the potential range of monetary losses caused by the closures.

(d) The marine disposal of waste materials can be controversial and difficult. Hansen *et al.* (1992) reported on how sites were selected in the New York Bight area of the USA Atlantic coast by evaluating candidate areas using GIS and available data on fisheries, currents, bottom sediments, slope angles, benthic productivity, and the existence of shipping lanes. It was even possible to predict the economic effect on fisheries of the designation of a disposal site, the potential for damaging species that have limited distributions in the areas, and the future growth and movement of sediments at a chosen dump site.

3.5 SOME POTENTIAL USES OF GIS IN MARINE FISHERIES SCIENCE

It is already clear that a single GIS package can perform a wide variety of management-related functions. These include the monitoring of change, comparative studies in the spatial and temporal contexts, the modeling of 'what if' scenarios, plus other graphical and quantitative capabilities. Additionally, GIS allows all analyses to be carried out at varying scales. This is particularly important to an activity such as fisheries where a target for management, such as for instance various species of salmon, may have lifestyle facets varying from spawning redds at a micro scale to international migration routes at a macro scale.

Meaden (1993) identifies seven database areas which must be of broad relevance to any marine fisheries GIS. Rather than outline these again we concentrate here on three broad database areas which are most directly relevant to fisheries research, i.e. not examined are areas such as fisheries effort or fish marketing (or areas which were exemplified in Section 3.4.2). After discussing each area in some detail, examples of studies are provided in which GIS might have proved of value. The aim is to be suggestive rather than speculative. The database areas identified are simply a means of classification into a range of spatio-related ideas, and within each there will be a huge range of potential subjects. GIS will additionally allow for cross-border analyses between the various areas.

3.5.1 Marine water conditions and habitats

Water qualitative parameters on which data are routinely collected include temperature, salinity, dissolved oxygen (DO) levels, water color/algae content, turbidity, sea currents, density, etc. Spatially variable readings already form the basis of some isoline maps. GIS storage and manipulative functionality would allow for the easy production of, for instance, spatial correlation or time-series analyses between and within variables. RS and acoustic SONAR techniques are allowing for the buildup of huge datasets on sea-bottom typology. Again, GIS will make it possible to map this information as well as constructing digital terrain models (DTM), thereby allowing for the visualizing of relationships such as between depth and various habitat types. Although there is abundant evidence of the importance of water qualitative factors to fish growth and recruitment, few detailed longer-term spatially based analyses have previously been possible.

The first example concerns the massive effects of environmental degradation on marine fisheries and the converse, the effects of fishing activities on marine environments. In the light of this, Hey (1992) sets out the case for far more stringent two-way spatial management and legal systems. GIS would provide an obvious means of implementing part of these measures. GIS would also have been useful in the type of modeling study carried out by Werner *et al.* (1993). As Figure 3.9 shows, the authors attempted to simulate, by the use of passive particles, the fate of the early-life stages of cod or haddock which spawn on the Georges Bank in the north west Atlantic. Any type of modeling which involves diffusion processes can be comparatively easily simulated. So, for instance, Thomson *et al.* (1992) proposed a conceptual model to show the effect of the North Pacific currents on sockeye salmon (*Oncorhynchus nerka*) migration routes. As inferred in their paper, this could similarly have been expedited by the use of GIS. A study by D'Amours (1993) sought to show the relationship between the distribution of cod, water temperature and oxygen levels in the Gulf of St. Lawrence. Figure 3.10 reveals how GIS could have been applied to advantage to achieve mapped interpolated surfaces for spatial auto-correlation analysis. In his recent oceanographic study of the North Sea, Huthnance (1991) provided a large number of maps to show physical relationships between aspects such as meteorology, tides, wave surges, sea level trends, circulation, etc. He concluded that for improved modeling we require better representations of reality. Finally, examples of gathering data for, and the potential for, bathymetric mapping with the aid of GIS are given in Mills and Perry (1992) and Somers (1992).

3.5.2 Natural marine resources

This section is essentially concerned with mapping estimates of marine biomass. Separate databases can be established for any particular geographic areas, and/or times, covering criteria such as individual fish species, sea mammals, plankton, cephalopods, shell fish, crustaceans, etc. Obviously these could be further subdivided into genera, subspecies, etc.

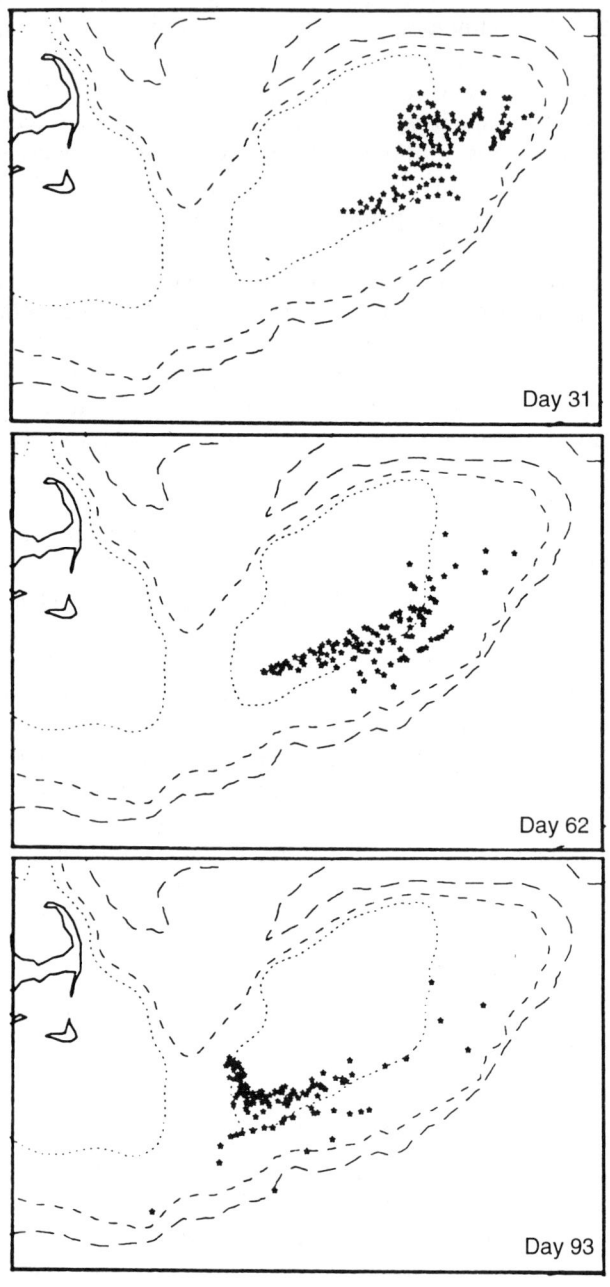

Most previous mapping of species distribution has simply utilized some sort of quantitative location symbol, such as proportional circles, or various classes of shading to illustrate species density variations. With specialized GIS algorithms it will be possible to refine distribution maps so that they show both species density variations and statistical variance. When mapping sampled data, it is important to indicate variance from the sample mean, especially since the objects being mapped are mobile and since their distributions may differ temporally as well as spatially. Plate 1 gives an idea of such a map.

Studies which exemplify the mapping of marine resources are often simple accounts which seek only to show specific species' distribution and perhaps density. Obviously, the potential for the use here of GIS is straightforward. Other studies introduce various relationships into these surveys. Shuntov *et al.* (1990) related fish biomass to the abundance of macroplankton in the Sea of Okhotsk (Figure 3.11); Iversen *et al.* (1993) showed the relationship between anchovy and sea surface temperatures in the Yellow Sea (Figure 3.12); Ware and Thomson (1991) showed the link between the long term variability of upwellings with total fish production in the north east Pacific Ocean; similarly Moita (1993) showed the spatial variability of phytoplankton communities in the upwelling off Portugal. GIS would not only have rapidly produced all of the maps used in these studies, but also would have provided a greater range of precision and opportunities for examining the data. Similarly, GIS would allow for greatly enhanced resolution of mapping if applied in studies such as Swain and Wade (1993), which showed the effect of population size on the geographic distribution of Atlantic cod (*Gadus morhua*) in the southern Gulf of St. Lawrence.

Besides being of use in the actual plotting of biomass distribution, GIS functionality can be applied to various statistical and resource modeling situations. Thus Simard *et al.* (1992) investigated the possibilities of using various statistical procedures to deal with estimation problems in the spatial autocorrelation of shrimp samples in the Gulf of St. Lawrence. Since they had very detailed data, this type of complex work would have been relatively straightforward given an appropriately sophisticated GIS package. Indeed, the maps they produced might well have been produced using a simple GIS (Figure 3.13). Crawford and Fox (1992) went to a great deal of trouble to produce 'visualizations' of echo-sounding data on detected fish biomass, using a graphics program SURFER to obtain the output. This could readily have been accomplished on a basic vector-based GIS package.

Fig. 3.9 Simulation showing monthly horizontal locations of passive particles released at 50 m depth above Georges Bank, north west Atlantic (after Werner *et al.*, 1993).

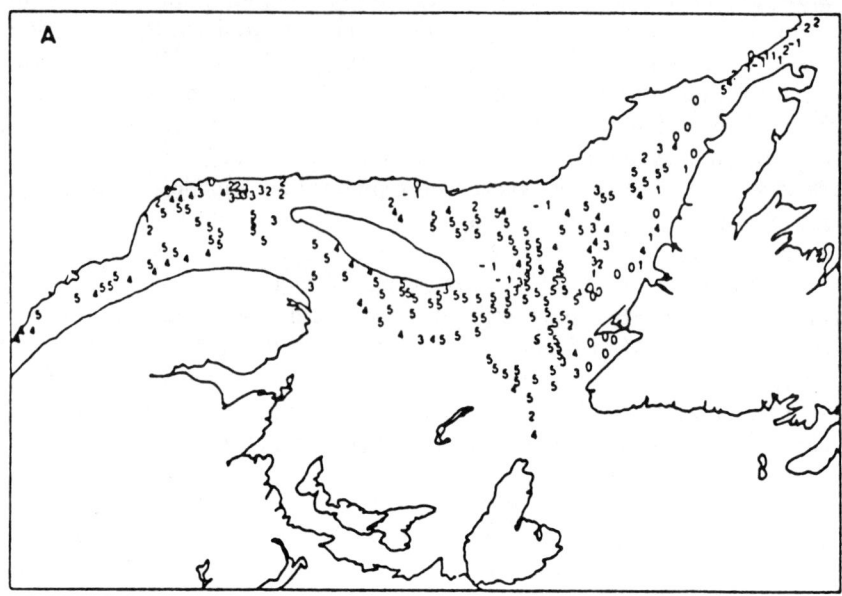

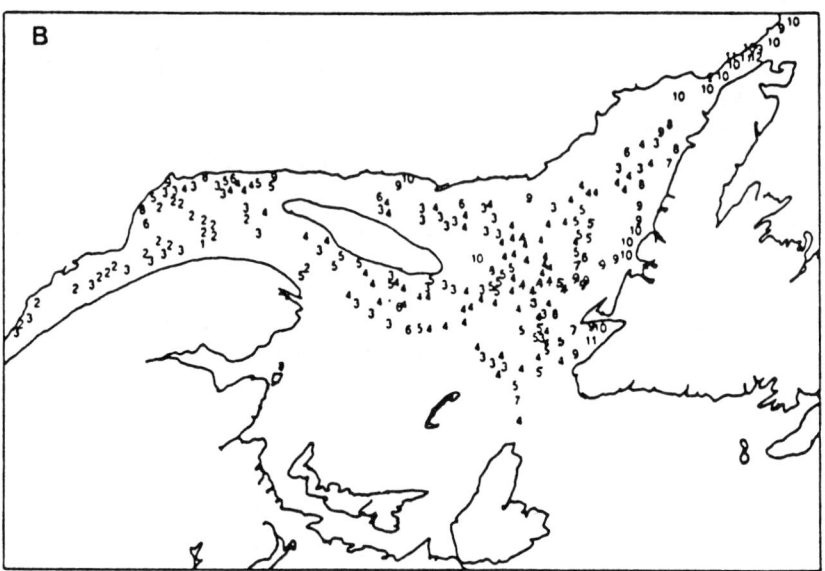

Fig. 3.10 Hydrographic station values for (A) bottom temperature (°C) and (B) bottom oxygen level ($mg\,l^{-1}$) as acquired during a survey in August–September, 1991 in the northern Gulf of St. Lawrence (after D'Amours, 1993).

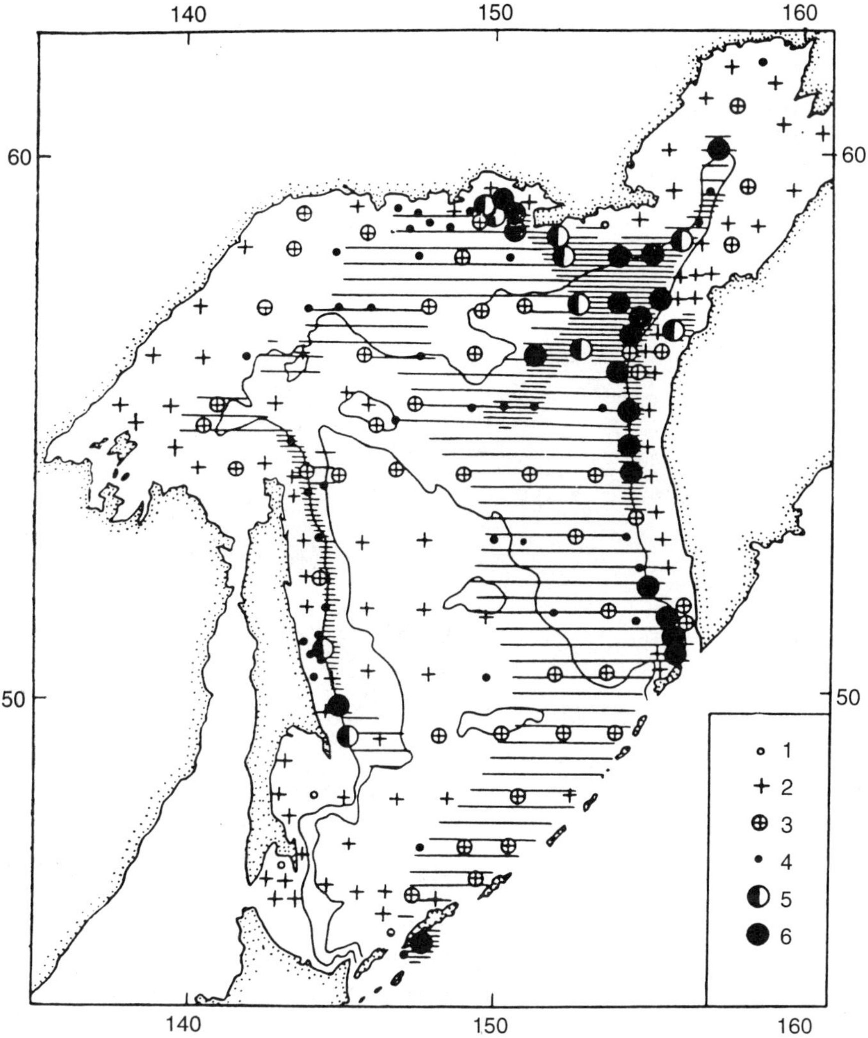

Fig. 3.11 Conventional map showing fish densities in epipelagic waters in the Sea of Okhotsk, June and July 1988, in tonnes per hour of trawling; 1, no catch; 2, less than 0.3; 3, 0.3–1.0; 4, 1.0–5.0; 5, 5.0–10.0; 6, over 10.0. Horizontal lines show echo soundings of fish, with the line density reflecting the intensity of the re-cordings (after Shuntov *et al.*, 1990).

3.5.3 Fisheries management and regulation

This section is largely concerned with GIS use for resource allocation. The need to manage fisheries in a spatial context has been well documented; Hinds (1992) gives an overview. One spatial aspect of fisheries regulation is

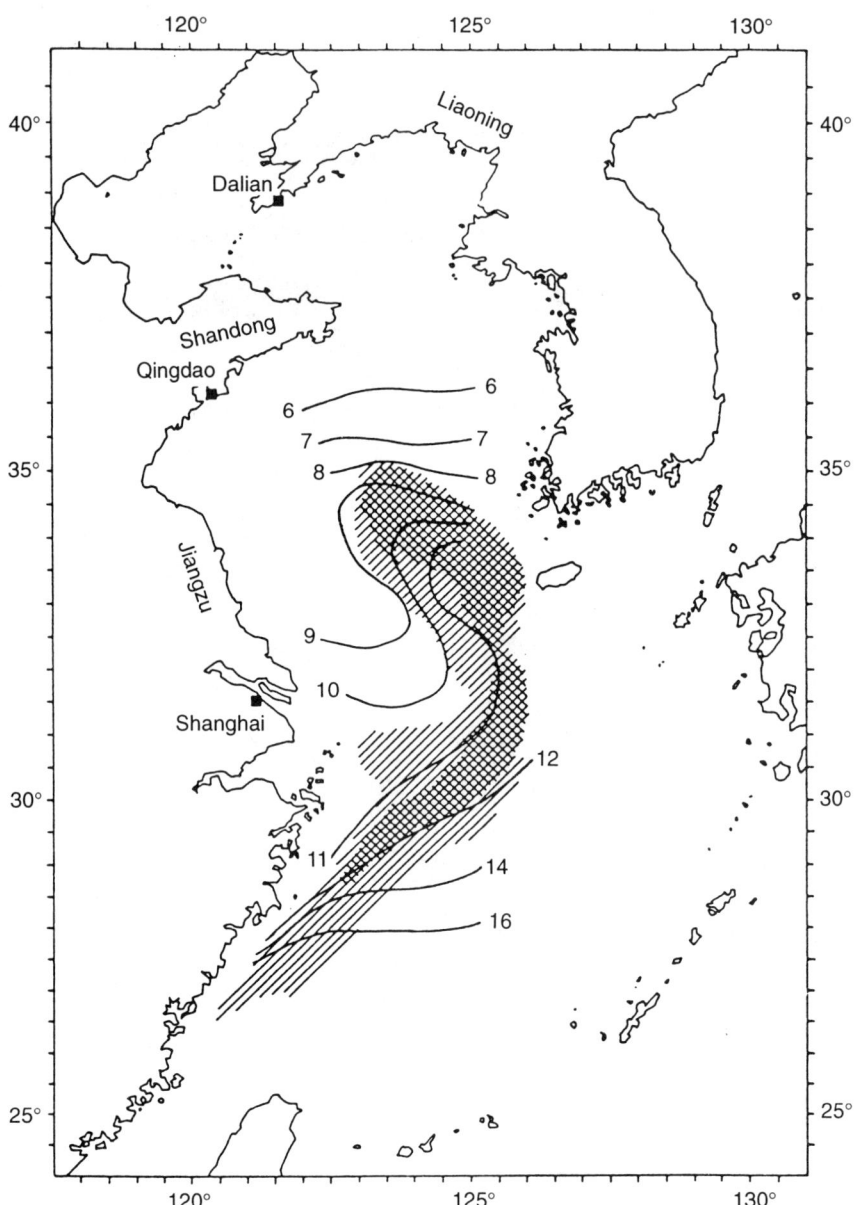

Fig. 3.12 Conventional map showing surface water temperatures (°C) and the generalized distribution of anchovy (*Engraulis japonicus*) during March in the Yellow and east China Seas for the period 1985–1990 (after Iversen *et al.*, 1993).

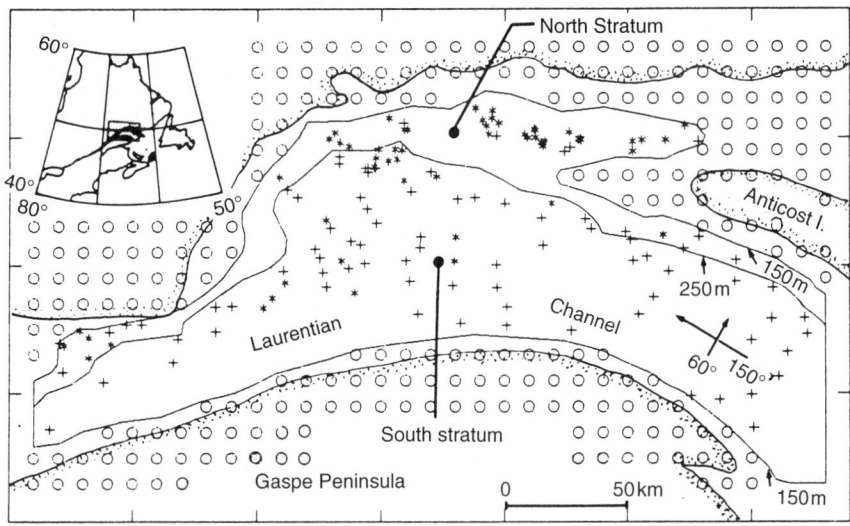

Fig. 3.13 Map showing sampling of northern shrimp (*Pandalus borealis*) in the western Gulf of St. Lawrence, August to October 1989: *, >1000 kg shrimp km⁻²; +, <1000 kg shrimp km⁻²; ○, areas outside the study region (after Simard *et al.*, 1992).

manifest from a legal viewpoint in the 200 mile Exclusive Economic Zone (EEZ); this might form the mapping boundary for any national fisheries GIS. Within this boundary, for both management and regulation purposes, it will be necessary to subdivide the area, possibly using a hierarchical nested cell structure, i.e. a more refined version of the ICES rectangles used in the North Sea. Variables plotted per cell might include quotas, fishing rights, total allowable catches (TACs), and indices of fish availability such as maximum sustainable yield or even restocking per cell.

An excellent example of how GIS might be employed in delimiting access rights is suggested by Bidi (1993) with reference to the Gulf of Guinea coastal states in West Africa. He suggests that the twelve coastal nations from Mauritania in the west to Nigeria in the east will need to have a degree of 'flexible cooperation' in their approach to managing the limited fish stocks of the area. In the neighboring area, i.e. between Guinea Bissau and southern Morocco, management of the prolific fish stocks urgently needs to be strengthened because of the almost total lack of control on access rights (Goffinet, 1992). The granting of territorial use rights for small-scale fisherfolk in almost any country, under community-based management schemes, is seen as a major way of enhancing the livelihood of the people involved (Ruddle, 1987). These three examples all involve rights of access and the right to determine levels of fishing effort and catches with the necessary legal backing. The management of such schemes would usefully employ GIS techniques at micro or macro scales, allowing for GIS learning in conditions where data gathering is unlikely to be too overwhelming. In the small and heavily overfished North Sea, Symes (1992, p. 336) recognizes that 'effective central control is essential' and that this must incorporate the need to vary the system of regulation to suit ever-changing conditions of stock availability and consequent total allowable catches. Similarly, in the Peruvian and Chilean upwelling coastal waters, which have witnessed huge stock fluctuations due to both natural and human intervention, Caviedes and Fik (1993) advocate strict resource management and restrictions on fish capture. In both of these latter studies, GIS could address all aspects of management.

3.6 WAYS OF INTRODUCING A FISHERIES GIS

The concern here is not with practical GIS implementation procedures and considerations, but with a broader look at the operative levels and ways in which a fisheries GIS might function. Clearly there is no right or wrong way of introducing a fisheries GIS and, within any country or region, GIS adoption approaches might vary according to specific aims and objectives and the different organizations which may be running the GIS. Thus it is easy to envisage that a fisheries GIS would perform a very different function for a private fishing operation than for a fisheries research institute.

3.6.1 Expanding database and functionality approach

This model of adoption is largely based upon the work being carried out on marine databases, specifically marine resource atlases (Ramster, 1991; 1994). Worldwide there has been a profusion of marine atlases. Until the 1980s these were largely in hard-copy format, but with the rapid adoption of computing technologies, digital formats are now favored. Ramster (1994) discusses developments of such atlases covering a number of coastal areas in North America and western Europe. An example is the United Kingdom Digital Atlas Project (UKDMAP). This currently comprises 462 maps of the seas around Britain which can be displayed in color on a low-cost PC. It has a simple menu interface, which allows for various map layers to be called up and integrated on the screen, plus a large volume of textual information on all of the databases. At its present stage of development, UKDMAP is more accurately described as an atlas than a marine GIS, since its only interactive functionalities (apart from overlaying) are straight-line distance measurement, zooming, and a cursor readout of latitude and longitude coordinates. However, it is envisaged that this functionality will slowly be extended, and there are already plans to provide UKDMAP users the ability to input their own data files and to edit existing maps. When this capability is adopted, then the atlas would be classed as a GIS, i.e. a reflection of the move towards multimedia.

3.6.2 Incremental evolution approach

An evolutionary approach is the one presently being taken in the majority of countries or regions exploring GIS. As previously described many GIS initiatives are underway, in an array of mostly public sector institutions, most of which are only peripherally associated with fisheries GIS. There are also a large number of often uncoordinated projects, each pursuing their own creditable aims, which may be establishing marine databases for motives unconnected with GIS. Given some form of georeferencing, these databases may later form the basis for a GIS. In this evolutionary approach there is no deliberate intention to have a full blown fisheries GIS, i.e. there are no aims and objectives which include working towards the management and control of fisheries *per se*. Eventually, however, an enormous number of datasets are likely to emerge, which could facilitate, if the data were suitably structured, the capability to have some kind of regional or national fisheries GIS.

3.6.3 Complete instigation approach

The purposeful instigation of a complete fisheries GIS has been described elsewhere (Meaden, 1993). This approach has as its overall aim the management of fisheries for the purpose of satisfying some predetermined objective such as to maximize the supply of fish or to bolster the recruitment rates for particular species. Clearly this system would be satisfying the demands of a government program, although it could be subcontracted for operative purposes to the private sector. The GIS might be instigated at a local, regional, national or even transnational level. Figure 3.14 gives an indication of some of the main linkages which would be necessary for a national GIS to be made operative in the United Kingdom.

The approach demands that systems implementation follow a careful plan. This would cover aspects such as data requirements, data formats and structure, the nature of database linkages, GIS functionality requirements, systems location, systems architecture, technical and other support, and the types of output required. In addition to these systems' operating requirements, the approach would give primary consideration to data gathering. This would clearly involve seeking data sources, establishing various kinds of communications linkages, and inevitably the setting up of data-gathering systems. Even though this is described as a 'complete'

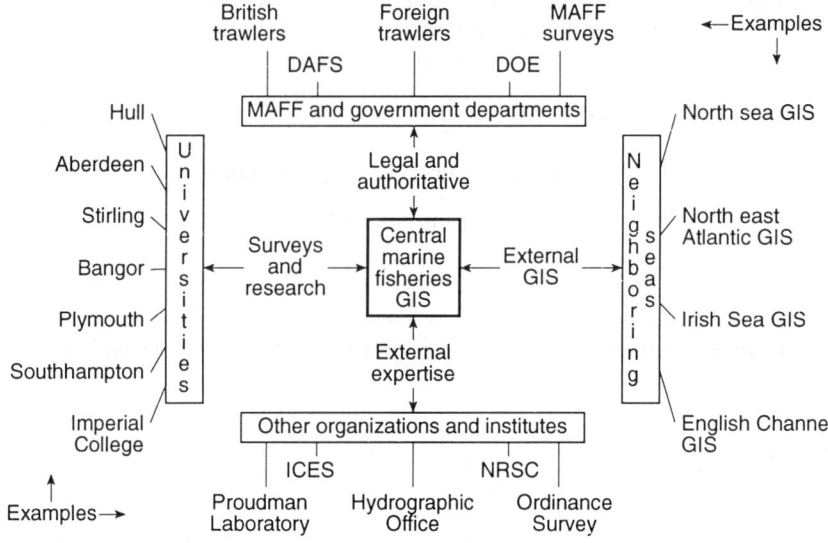

Fig. 3.14 The main linkages in a possible central marine fisheries GIS for the UK: MAFF, Ministry of Agriculture, Fisheries and Food; DAFS, Department of Agriculture and Fisheries, Scotland; DOE, Department of Environment; ICES, International Council for Exploration of the Sea; NRSC, National Remote Sensing Centre.

approach, output from the system would emerge only incrementally, perhaps in areas seen as being of particular importance or in which reliable output could be more easily obtained. It would invariably take some time before this GIS was capable of delivering all of the system's potential or all of the management requirements.

3.7 CONCLUSIONS AND CHALLENGES

Fishery modeling has been slow to develop a functional or quantitative awareness of population geography. Once a 'unit stock' has been defined, geography is often ignored except with regard to management options involving area closures. Indeed geography has not been easy to incorporate into useful, practical models of population dynamics. (MacCall, 1990, p. 120)

MacCall's thesis was to show the urgent need of subsuming geographic thinking into the modeling and management of fisheries. Other organizations, such as the Food and Agriculture Organisation of the UN (Caddy and

Garcia, 1986), have also long recommended that spatial management via mapping should be a prerequisite for the strategic development of fisheries. Although this has been difficult in the past, the advent of GIS has opened up extraordinary opportunities, many of which we have described.

From both the developmental and applications perspectives, GIS has made significant advances, especially during the 1990s. The management of a range of public and private enterprises has benefited from having a graphical/mapping perspective on problems associated with their spatial organization. Likewise GIS is being used in a wide range of research activities for both monitoring and modeling. However, despite all the potential advantages, the penetration of GIS into fisheries-related areas has been slow. This implies that a number of challenges have to be met before GIS becomes more universally accepted. The main challenges are seen to be the following:

(a) The adoption of 3-D data collection structures and formats into GIS software or into database management software. This needs to be carried out in conjunction with continuing research into visualization and presentation of 3-D GIS imagery.
(b) The improvement in data gathering methodologies, especially with regard to regularizing unit data cell areas and to the obligation of fishing vessels to record and georeference fishing catch and effort data.
(c) Continuing geostatistical research methods, plus their applications to GIS software, so that the best algorithms can be adopted for obtaining and applying biomass estimates within unit areas. Additional research is also needed into the integration of dynamic modeling into GIS and into the optimum ways of mapping statistical variance.
(d) Persuading leading governmental fisheries institutions to take a firm lead in GIS implementation with a view to setting aims and objectives for a central fisheries GIS, organizing its implementation, ensuring inter-governmental cooperation, standardizing formats and procedures for data collection, and ensuring that systems linkages are set up and maintained.
(e) Examining ways in which GIS functionality and hardware systems could best be installed and operated on both research and commercial fishing vessels.

Given that the subject area of fisheries is so vast, GIS adoption is going to be incremental in terms of both systems set up and priorities for database introduction. We see no reason why certain databases cannot be quickly and efficiently introduced into a central fisheries GIS, even if for the present they may be in a very rudimentary form. The fisheries community also knows that certain macro and micro marine areas are in a perilous state with regard to their fisheries future. Given this predicament, the powerful management systems offered by GIS should be both widely and urgently adopted.

REFERENCES

Allan, T.D. (1992) The marine environment. *International Journal of Remote Sensing,* **13**(6,7), 1261–76.

Alonso, W. (1960) A theory of the urban land market. Papers and Proceedings, Regional Science Association, (6), 149–57.

Bernhardsen, T. (1992) *Geographic Information Systems,* VIAK IT, Arendal, Norway.

Beveridge, M.C.M. and Ross, L.G. (1991) *Environment, site selection and planning: The role of geographic information systems in aquaculture,* IFS Regional Workshop on Ecology of Marine Aquaculture, Osorno, Chile, 18–23 November, 1991.

Bidi, J.T. (1993) Geographical approach to the exclusive economic zone in Guinea Gulf. *Ocean & Coastal Management,* **19**, 137–55.

Bostock, J. (1994) *Fishing for information: A guide to network and on-line resources in aquaculture and aquatic science,* Institute of Aquaculture, University of Stirling, Scotland.

Butler, M.J.A. and LeBlanc, C. (1990) *Marine information networks: Challenge for the 90s,* GIS for the 1990s – Proceedings, National Conference, March 5–8, Ottawa, Canada, 1197–252.

CCTA (1993) *Geographic Information Systems: A Buyers Guide,* HMSO, London, England.

Caddy, J.F. and Garcia, S. (1986) Fisheries thematic mapping: A prerequisite for intelligent management and development of fisheries. *Oceanographie Tropicale,* **21**(1), 31–52.

Caviedes, C.N. and Fik, T.J. (1993) Modelling change in the Peruvian–Chilean eastern Pacific fisheries. *GeoJournal,* **30**(4), 369–80.

Collins, N. and Hurlbut, S. (1993) *Environmental risk analysis of salvaging the Irving Whale,* GIS '93, March 23–25; Ottawa, Canada.

Crawford, R.E. and Fox, J.K. (1992) *Visualization of echo sounder data with a micro computer,* Canadian Technical Report on Fishery & Aquatic Science, No. 1840, pp. iv–17.

D'Amours, D. (1993) The distribution of cod (*Gadus morhua*) in relation to temperature and oxygen level in the Gulf of St. Lawrence. *Fisheries Oceanography,* **2**(1), 24–9.

Gilbert, C. (1994) Portable GPS for mapping: Features versus benefits. *Mapping Awareness,* **8**(2), 26–9.

Goffinet, T. (1992) Development and fisheries management: The case of northwest Africa. *Ocean & Coastal Management,* **17**, 105–36.

Hansen, W.J., Goldsmith, V., Clarke, K.C. and Bokuniewicz, H. (1992) A marine GIS examines dredging and wetland operations in the New York Bight. *Geo Info Systems,* **2**(4), 52–6.

Harper, B. and Curtis, M. (1993) Coastal zone mapping. *Mapping Awareness & GIS in Europe,* **7**(1), 17–19.

Hey, E. (1992) A healthy North Sea ecosystem and a healthy North Sea fishery: Two sides of the same regulation? *Ocean Development and International Law,* **23**, 217–38.

Hinds, L. (1992) World marine fisheries: Management and development problems. *Marine Policy,* **16**(5), 394–403.

Huthnance, J.M. (1991) Physical oceanography of the North Sea. *Ocean & Shoreline Management,* **16,** 199–231.

Iversen, S.A., Zhu, D., Johannessen, A. and Toresen, R. (1993) Stock size, distribution and biology of anchovy in the Yellow Sea and East China Sea. *Fisheries Research,* **16,** 147–63.

Kam, S.P., Paw, J.N. and Loo, M. (1992) The use of remote sensing and geographic information systems in coastal zone management, in *Integrative Framework and Methods for Coastal Area Management,* (eds Chua, T.-E. and Scura, L.F.), ICLARM Conf. Proc. 37: pp. 107–31.

Kapetsky, J.M., Hill, J.M. and Dorsey Worthy, L. (1988) A geographical information system for catfish farming development. *Aquaculture,* **68,** 311–20.

Kapetsky, J.M., McGregor, L. and Nanne, H.E. (1987) *A geographical information system to plan for aquaculture, A FAO–UNEP/GRID study in Costa Rica,* FAO Fisheries Technical Paper No. 287, FAO, Rome.

Knott, J. and Shiers, D. (1994) Beam me a map, Scotty! Mobile maps make a GIS portable. *Mapping Awareness,* **8**(2), 32–4.

Kolacny, A. (1968) *Cartographic Information – A Fundamental Notion and Term in Modern Cartography,* Czechoslovak Committee on Cartography, Prague, Czechoslovakia.

Legault, J.A. (1990) *Using a geographic information system to evaluate the effects of shellfish closure zones on shellfish leases, aquaculture and habitat availability,* Canadian Technical Report on Fishery & Aquatic Science, No. 1882E, pp. iv–10.

Li, R. and Saxena, N.K. (1993) Development of an integrated marine geographic information system. *Marine Geodesy,* **16,** 293–307.

Liebig, W. (1994) Protecting the environment: GIS and the Wadden Sea. *GIS Europe,* **3**(2), 34–6.

Lo, C.P. and Hutchinson, W.T. (1991) Determination of turbidity patterns of the Zhujiang estuarine region, South China, using satellite images and a GIS approach. *Geocarto International,* **3,** 27–38.

MacCall, A.D. (1990) *Dynamic Geography of Marine Fish Populations,* University of Washington Press, Washington, USA.

Maussel, P.W., Everitt, J.H., Escobar, D.E. and King, D.J. (1992) Airborne videography; Current status and future perspectives. *Photogrammetric Engineering and Remote Sensing,* **58,** 1189–95.

Meaden, G.J. (1987) Where should trout farms be in Britain? *Fish Farmer,* **10**(2), 33–5.

Meaden, G.J. (1993) *Instigation of the world's first marine fisheries GIS,* ICES Statutory Meeting, No. C.M. 1993/D:64, Dublin, Ireland.

Meaden, G.J. (1994) The one that got away? GIS in marine fisheries. *Mapping Awareness,* **8**(7), 20–3.

Meaden, G.J. and Kapetsky, J.M. (1991) *Geographical information systems and remote sensing in inland fisheries and aquaculture.* FAO Fisheries Technical Paper. No. 318, FAO, Rome.

Mills, G.B. and Perry, R.B. (1992) EEZ bathymetric mapping for ocean resource management. *Sea Technology,* **33**(6), 27–36.

Moita, M.T. (1993) *Spatial variability of phytoplankton communities in the upwelling region off Portugal,* ICES Statutory Conference, No. C.M. 1993/L:64, Dublin, Ireland.

Paw, J.N., Diamante, D.A.D., Robles, N.A., Chua, T.E., Quitos, L.N. and Cargamento, A.G.A. (1992) *Site selection for brackishwater aquaculture development and mangrove reforestation in Lingayen Gulf, Philippines using geographic information systems,* Canadian Conference on GIS, 24–26 March 1992, Ottawa, Canada.

Price, W.F. (1992) Data acquisition for ground surveys. *Mapping Awareness & GIS in Europe*, **6**(6), 33–7.

Ramster, J.W. (1991) A proposed digital atlas of the resources of the North Sea and its potential value in planning. *Ocean & Shoreline Management*, **16**, 354–73.

Ramster, J.W. (1994) Marine resource atlases: The implications for policymakers and planners. *Underwater Technology*, **19**(4), 12–19.

Riddell, K.J. (1992) Geographic information and GIS: Keys to coastal zone management. *GIS Europe*, **1**(5), 22–5.

Robinson, A.H. and Petchenik, B.B. (1977) *The map as a communication system. Cartographica: The Nature of Cartographic Communication*, (ed. Guelke, L.), Monograph No. 19, British Cartographic Society, pp. 92–110.

Ross, L.G., Mendoza, Q-M, E.A. and Beveridge, M.C.M. (1991) *The use of geographical information systems for site selection for coastal aquaculture*, Institute of Aquaculture, University of Stirling, Scotland.

Ruddle, K. (1987) The management of coral reef fish resources in the Yaeyama Archipelago, southwestern Okinawa. *Galaxea*, **6**, 209–35.

Schauser, U.-H., Boedeker, D., Matusek, S. and Klug, H. (1992) *The use of a Geographic Information System for Wadden Sea conservation in Schleswig-Holstein*, Publication Series No. 20, Netherlands Institute for Sea Research, pp. 281–3.

Shuntov, V.P., Volkov, A.F., Abakumov, A.I., Shvydkiy, G.V., Temnykh, O.S., Vdovin, A.N., Starstev, A.H. and Shebanova, M.A. (1990) Composition and present status of the fish community in epipelagic waters of the Sea of Okhotsk. *Journal of Ichthyology*, **30**(4), 116–29.

Simard, Y., Legendre, P., Lavoie, G. and Marcotte, D. (1992) Mapping, estimating biomass, and optimising sampling programs for spatially autocorrelated data: Case study of the Northern Shrimp (*Pandalus borealis*). *Canadian Journal of Fishery & Aquatic Science*, **49**, 32–45.

Simpson, J.J. (1992) Remote sensing and geographical information systems: Their past, present and future use in global marine fisheries. *Fisheries Oceanography*, **1**(3), 238–80.

Somers, M.L. (1992) Progress in deep ocean mapping. *Underwater Systems Design*, **14**(3), 10–21.

Swain, D.P. and Wade, E.J. (1993) Density-dependent geographic distribution of Atlantic Cod (*Gadus morhua*) in the Southern Gulf of St. Lawrence. *Canadian Journal of Fishery & Aquatic Science*, **50**, 725–33.

Symes, D. (1992) The common fisheries policy and UK quota management. *Ocean & Coastal Management*, **18**, 319–38.

Thomson, K.A., Ingraham, W.J., Healey, M.C., Leblond, P.H., Groot, C. and Healey, C.G. (1992) The influence of ocean currents on latitude of landfall and migration speed of sockeye salmon returning to the Fraser River. *Fisheries Oceanography*, **1**(2), 163–79.

Tomlinson, R.F. (1989) Presidential address: Geographic Information Systems and geographers in the 1990s. *The Canadian Geographer*, **33**(4), 290–8.

Von Thunen, J.H. (1826) *Der isolierte staat in beziehung auf landwirtschaft und natio-nalokonomie*. English translation edited by P. Hall (1966) *Von Thunen's Isolated State*, Oxford, UK.

Wardle, D. (1992) Information systems in the UK Hydrographic Service. *Mapping Awareness & GIS in Europe*, **6**(7), 23–7.

Ware, D.M. and Thomson, R.E. (1991) Link between long-term variability in upwelling and fish production in the Northeast Pacific Ocean. *Canadian Journal of Fishery & Aquatic Science*, **48**, 2296–306.

Weber, A. (1909) *Theory of the location of industries*. English translation by C.J. Friedrich, Chicago University Press, Chicago, USA.

Welch, R., Remillard, M. and Alberts, J. (1992) Integration of GPS, remote sensing, and GIS techniques for coastal resource management. *Photogrammetric Engineering & Remote Sensing*, **58**(11), 1571–8.

Werner, F.E., Page, F.H., Lynch, D.R., Loder, J.W., Lough, R.G., Perry, R.I., Greenberg, D.A. and Sinclair, M.M. (1993) Influences of mean advection and simple behaviour on the distribution of cod and haddock early life stages on Georges Bank. *Fisheries Oceanography*, **2**(2), 43–64.

Chapter four

Quantitative fisheries research surveys, with special reference to computers

Kenneth G. Foote

4.1 INTRODUCTION

Fisheries research surveys are designed and conducted for a variety of purposes. Some examples are those of defining the geographical limits of distribution of a stock, studying the fisheries part of an ecosystem, and describing diurnal movements or seasonal migrations of a stock. The sole purpose considered here, however, is that of abundance estimation.

Surveys of fish stock abundance are generally carried out by researchers to avoid the well-known pitfalls of fishery-dependent methods. Four widely used generic surveys are described: egg and larval survey, mark–recapture experiment, trawl survey, and acoustic survey.

Common to each of these generic surveys, in addition to aim, is the use of computing resources. This varies from survey type to survey type, because of the kinds of operations that are performed, and also with respect to the particular survey, because of the number of data to be treated. Typical operations involving computers are enumerated:

(a) *Data collection* – the primary data are often gathered with instruments whose control and operation is rapid, repetitive, and complicated too. A dedicated processor, or special-build computer, is commonly used. The echo sounder is an example.
(b) *Data preprocessing* – the numbers of certain kinds of data, especially acoustic data, may be so large that preprocessing in real time is necessary. A special-purpose computer may be necessary.

Computers in Fisheries Research.
Edited by Bernard A. Megrey and Erlend Moksness.
Published in 1996 by Chapman & Hall, London. ISBN 0 412 59550 8

(c) *Data display* – in operating some common instruments in research surveys, such as trawl-monitoring instruments and echo sounders, real-time display of data is an end in itself. Important information is conveyed by the image. Presentation of this is generally effected with a dedicated processing unit.

(d) *Data logging* – survey data are often so numerous that the only rational way to store these during collection, if after preprocessing, is through the mass storage devices associated with a digital computer. Retrieval of data, as for postprocessing or analysis, is similarly effected by means of a digital computer.

(e) *Data postprocessing* – this may be necessary for one or more reasons. Quality control of collected data prior to their analysis may be expedited by assembly of frequency distributions, computation of statistics, or searching for data outliers, or extreme values, in addition to displaying data, each of which operations may be supported by a digital computer. Acoustic data may not be properly processed sometimes until after substantial numbers have been collected. In such cases, rapid post-processing may be necessary, for example, in retrieving and displaying echosounder data and in performing echo integration on subsets of these data as defined by an operator using interactive graphics applied to the echogram.

(f) *Data analysis* – derivation of estimates of abundance may involve relatively simple operations, as in egg and larval surveys or mark–recapture experiments, or more complicated operations, as in digesting measurements of fish density along line transects to yield an estimate of abundance over an area. Use of a digital computer undoubtedly simplifies conduct of the mathematical operations, but may also be necessary because of the sheer volume of basis data.

In every case where spatial data are involved, geostatistical analysis may be desired. The structure function is first computed, to characterize the observed spatial distribution by covariance function or variogram. This may then be used for interpolation, as in mapping the distribution where it is not directly measured. It may also be used in computing the estimation variance. Both mapping and computation of the estimation variance are computationally intensive operations. Geostatistics may be equally useful in egg and larval surveys, trawl surveys, and acoustic surveys.

A spectrum of demands on processor type and computing power is thus made by the four generic surveys. In certain respects these are quite different, for example, because of the differing numbers of data and differing degrees of complexity in analysis. At the same time, the computer in one form or another is generally recognized as essential to each survey type.

In the following, the generic surveys are reviewed in presumed historical order. The greatest weight is given to the newest survey type, the acoustic survey, which depends at every stage of its execution on the automatic control of instruments or on the processing of large numbers of data.

4.2 EGG AND LARVAL SURVEYS

Given sufficient information about reproduction biology and early life history, the abundance of the spawning component of a number of fish stocks may be estimated by an egg and larval survey. This involves sampling by physical capture of eggs and larvae over the region of occurrence, performance of laboratory measurements to determine biological quantities related to adult fecundity *inter alia*, and simple estimation of spawning population size by substitution of measured quantities in an equation.

Sampling requirements for performing a survey are:

(a) definition of the time of spawning and region of occurrence of eggs and larvae;
(b) adequate spatial and temporal sampling of this;
(c) quantitative knowledge of the volume associated with each individual sample.

It is essential that eggs and larvae of the target species, including development stage, be identified when analyzing samples. It must be possible to determine egg mortality and incubation rates, or stage duration. Knowledge of adult fecundity is assumed. Allowance for advection must generally be made.

Some representative studies aiming to elucidate critical factors for performing an egg and larval survey are presented in Ellertsen *et al.* (1989), Fossum and Moksness (1993), Hempel and Hempel (1971), Munk and Christensen (1990), Sundby (1991), Sundby *et al.* (1989), Talbot (1977), and Westgård (1989). In Fossum and Moksness (1993), for example, hatching period is estimated. The efficiency of gear in sampling larvae is specifically addressed in Munk (1988) with respect to herring (*Clupea harengus*) and in Suthers and Frank (1989) with respect to cod (*Gadus morhua*).

Population size is computed according to one of three methods. These depend on determination of annual egg production, daily reduction in fecundity, and daily egg production, respectively. Corresponding equations are quite simple. Details on the methods as well as examples of successful pelagic egg surveys and demersal egg surveys are given in Gunderson (1993). A particular example of the daily-egg-production method is given for Cape anchovy (*Engraulis capensis*) in Hampton *et al.* (1990). Total production of fertilized eggs has been both measured and calculated for cod and plaice (*Pleuronectes platessa*), establishing the usefulness of plaice egg surveys for stock assessment (Heessen and Rijnsdorp, 1989). Examples of larval abundance surveys are given for herring in Anthony and Waring (1980), Burd and Holford (1971), Hempel and Schnack (1971), Lough *et al.* (1985), Saville (1971), and Stevenson *et al.* (1989).

Data quantities are quite modest. However, as with other survey types involving spatial sampling, a geostatistical analysis (Petitgas, 1993) may be computationally intensive. Particular aims of such an analysis may be determination of structure through a generally anisotropic covariance function or variogram, and use of this in mapping the distribution of eggs and larvae and in estimating the variance of the mean abundance estimate due to coverage of the region of occurrence in relation to observed distributional properties.

As a future development, information on preferences for spawning site or oceanographic conditions might be used in a deterministic drift model together with wind and current data to predict egg and larval distribution, as has been done with respect to cod eggs and larvae on Georges Bank (Lough *et al.*, 1994) and cod larvae in the Barents Sea (Ådlandsvik and Sundby, 1994). This might be exploited in allocating sampling effort. It might also be used with geostatistical models to improve distribution maps and the estimation of variance attached to the abundance estimate. Such modeling work is generally computationally intensive.

4.3 MARK–RECAPTURE EXPERIMENTS

Mark–recapture experiments on fish have an illustrious historical precedent on humans, which marks the beginning of demographic statistics, cited by Ricker (1975). In applications to fish, such experiments are typically used to determine exploitation rate as well as total fish population.

In essence, a number of fish are marked, as by tagging, and released. After allowing a suitable time for mixing, the subsequent fish catch is registered and number of recaptures identified. Under certain simplifying assumptions, the population abundance is the ratio of total catch to rate of exploitation, where this rate is estimated by the ratio of the number of recaptures to number of marked fish. This so-called Petersen method is described in detail by, among others, Ricker (1948) and Seber (1973).

Ricker and Seber each enumerates six assumptions that must be fulfilled for application of the mentioned, simple formula. These are essentially the same, differing principally in the degree of explicitness of the assumptions of mortality (Ricker, 1948) and equal sampling probabilities in first and second samples (Seber, 1973). In accordance with the summary advice given in Gunderson (1993) on biological aspects of survey design, availability, sampling gear, and vulnerability and selectivity, these assumptions should be verified. For the Petersen method, studies would have to determine the following (Ricker, 1975):

(a) natural mortality due to marking, i.e. marking mortality;
(b) degree to which marks are lost;
(c) vulnerability of marked fish to catching;
(d) degree to which mixing of marked fish with the population of unmarked fish is uniformly random, or degree to which fishing effect is proportional to density;
(e) degree to which marks are detected among recaptured fish;
(f) degree to which recruitment to the catchable population during the time of recapture is negligible.

Effects of violations in the assumptions are considered by the cited authors, as are more general cases of inference about fish populations. In particular, the following generic cases are considered: constant and variable survival rates, closed populations with both single-mark release and multiple releasing, and open populations with mark release before or during the sampling period.

Jones (1976) gives a very accessible, detailed account of the use of mark–recapture data for abundance estimation, as well as for determination of survival rates, growth, and movement parameters. Associated mathematics are not computationally demanding, but the number of marks recovered in large-scale experiments, together with the desire to exploit information on recapture site and biological parameters of recaptured fish, make the digital computer an essential tool in the analysis of such data.

Practical examples on the role of mark–recapture experiments in fish stock abundance estimation are found in, for example, Aasen (1958), Dragesund and Jakobsson (1963), and Dragesund and Haraldsvik (1968). Exemplary detailed studies on the tagging or marking process itself are found in, for example, Nakashima and Winters (1984), Parker *et al.* (1990), and Wetherall (1982).

4.4 TRAWL SURVEYS

Trawl surveys aim to estimate abundance over a specific geographical area by means of a series of trawl hauls. Trawl locations may be distributed over the region according to a number of patterns. Basic examples are uniform or random grids, both with and without stratification.

The major concern in nearly every trawl survey is that of representativity of sampling. Determining how representative trawl samples are involves adjunct studies that often require more effort than that of the survey itself. Such studies examine the phenomena of size selectivity, vulnerability, and fish behavior, including avoidance reactions to vessel and gear. As these generally depend on species, size, biological state, season, depth, and light level, among other influences, quantifying the fundamental sampling process is indeed a formidable undertaking. Exemplary studies focusing on some of the mentioned problems for bottom trawls are described in Engås and Godø (1986, 1989a,b), Godø and Engås (1989), Godø and Sunnanå (1992), Godø and Walsh (1992), Harden Jones *et al.* (1977), and Ona and Godø (1990). The problems of changes in vertical and horizontal distributions are addressed in Godø and Wespestad (1993), for example. The cited study of Godø and Walsh (1992) has led to a change in trawl type, from rubber bobbin ground gear to heavy rockhopper ground gear.

Under certain conditions, trawl surveys can yield a relative measure of abundance that may serve as an index. If the trawl performance can additionally be quantified, so that the effective area swept by a bottom trawl or effective volume swept by a pelagic trawl is known for each individual haul, and the corresponding efficiency is known, then a trawl survey should be capable of yielding absolute estimates of abundance. Godø (1994) summarizes factors affecting the reliability of abundance estimates of bottom fish.

Vital data on trawl performance is provided by such pioneering cable-free instrumentation as that produced by SCANMAR AS, for example, the SCANMAR System 400. This includes sensors that transmit data on gear depth, trawl velocity including sinking or rising rates, distance between trawl doors or wings, height of trawl opening, temperature at operating depth, and cod-end filling among other things. The so-called trawleye is also part of the system; it monitors fish concentration in the vicinity of the trawl opening, as well as bottom contact, providing data for real-time display in echogram form on a color graphics monitor.

The SIMRAD Integrated Trawl Instrumentation (ITI) system provides similar information on trawl performance by cable-free instruments using an acoustic link. This is used in both directions, however, as the instruments operate only on command, when interrogated, thus prolonging battery use between recharging periods.

Trawl surveys have traditionally been directed at bottom fish, but may apply equally well to pelagic fish, given suitable conditions. Certainly trawling of pelagic fish is important in acoustic surveys, hence the expressed concern about representativity in sampling by both kinds of trawling gear. In the case of pelagic trawls, the crucial issue of capture efficiency is described in Hylen *et al.* (1995). This is discussed further with respect to an acoustic index (Haug and Nakken, 1977) and a catch-rate index (Randa, 1984) in Nakken *et al.* (1995). Astthorsen *et al.* (1994) describe the derivation of abundance indices for juvenile cod by pelagic trawling and present results as a time series over the period 1970–1992.

Just as the design of trawl surveys may vary from uniform sampling regimes to those seeking randomness, so do the precise analyses of trawl survey data vary. When direct use is made of the spatial location of trawls, by means of geostatistics, computations may require the use of a powerful digital computer. Results may be expressed in terms of structure functions, maps of the distribution, and global estimate of abundance with variance estimate. When spatial information is ignored, computation of total abundance may merely involve summing catch numbers, possibly weighted by the respective stratum area or by the total survey area. Given the number of strata sampled in some surveys, use of a digital computer is a practical necessity.

During the International Workshop on Survey Trawl Mensuration held in St. John's, Newfoundland, in 1991 (Walsh *et al.*, 1993), 76 factors were listed as influencing survey trawl performance and fish capture efficiency. Should a sufficient number of the more important of these be quantified in the future and models for their compensation be developed, then this information might be used to improve local estimates of fish density. Abundance estimates over an area would be correspondingly improved. Application of a large number of factors and models would undoubtedly require a degree of computing power.

4.5 ACOUSTIC SURVEYS

Quantitative acoustic surveys are relatively new compared to the generic surveys already discussed, depending as they do on both sophisticated electromechanical devices (called transducers), electrical circuitry, electronic controls, and electronic processors. Nonetheless, development of these has paralleled general developments in materials and electronics technology. The possibility of performing rapid and remote sampling of fish populations on the synoptic scale has been a powerful attraction for new applications in fisheries research. One of these, that of abundance estimation, is now reviewed.

4.5.1 Generic method

Prerequisites for conventional acoustic surveys are a transducer or sonar, electronics for controlling transmission and reception, a display device, a platform to bear these, namely a vessel or towed vehicle, and a digital computer. Essential functions of the instruments are those of echo sounding, data display, data processing, and data storage. The associated echo sounder or sonar system is generally operated in a calibrated state. Details on acoustic and data processing systems are given below.

The survey area is spanned by a network of transects. A major aim of the design of this is acquisition of maximal information about the spatial distribution of fish. Typical constraints are the known or suspected boundaries of the stock, possible large-scale movements of the fish during the course of the survey, diurnal patterns of fish movement, availability of ship time, possible need to conduct trawling or perform hydrographic or other oceanographic measurements concurrently with the acoustic survey, among other considerations. *A priori* information about the fish stock is always valuable, as in allocation of sampling effort by stratification. Current practice is reviewed by Simmonds *et al.* (1992).

In a particular case, that of Norwegian spring-spawning herring when wintering in a fjord system, the fish are generally surveyed in two or more stages. Initially, the fjord areas are covered by a large-scale zigzag design, shown in Figure 4.1. This aims to determine the location of significant quantities, in addition to providing the basis for the first abundance estimate. Identified subareas or significant strata are surveyed subsequently according to an *ad hoc* design that aims to maximize information about the particular concentrations. In December 1993, for example, the predominant concentration of herring was in central Ofotfjorden. Its second acoustic coverage was made in accordance with the design shown in Figure 4.2.

The most basic measured quantity is the echo signal recorded at a particular location. This contains information about the density of scatterers as a function of distance from the acoustic source and receiver, which are usually colocated, as assumed here. The intensity of the received signal, generally but not always processed with some form of range compensation, is displayed on paper or electronic screen. When successive echo signals are aligned, the resulting echogram presents an acoustic image of scatterers along the transect.

Quantitative processing of the same information may specify the acoustic density of particular echogram features, or scatterers, as a function of location, including depth in the case of ordinary vertically oriented echo-sounder beams or range in the case of sonar beams of general orientation. Because of the overall aim of stock abundance estimation, processing of echo-sounder data is performed in systematic fashion.

An important part of data postprocessing or analysis is allocation of the echo record to particular scatterer classes. In this, echo quantities are assigned to fish species and size categories, if possible. Auxiliary information useful in making the assignment includes composition of trawl hauls, data from oceanographic sensors, e.g. salinity-temperature-depth (STD) sonde, appearance of the echogram, and knowledge of fish biology. Results of

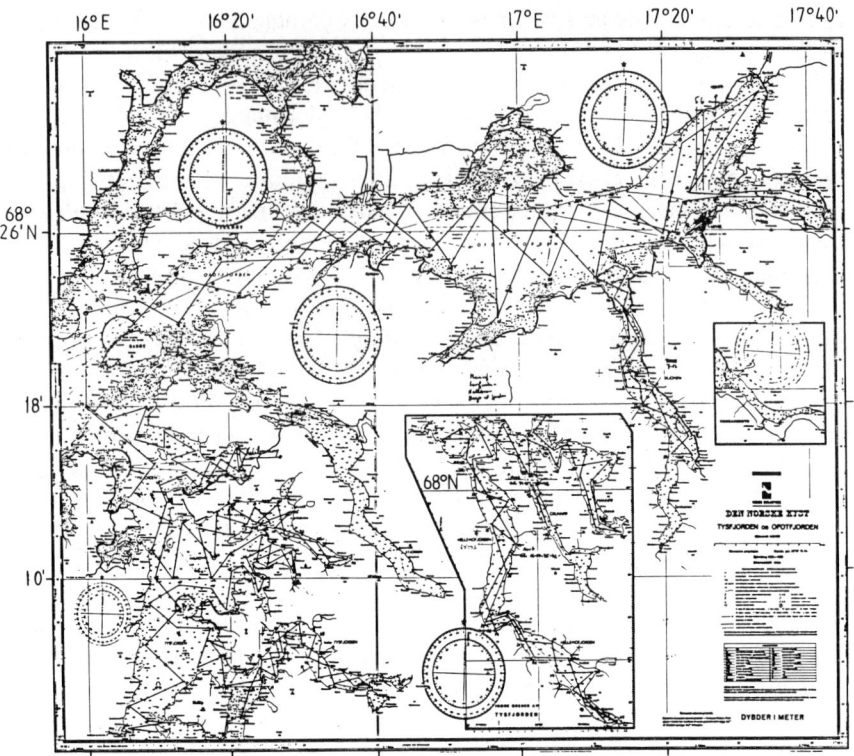

Fig. 4.1 Large-scale zigzag survey design applied initially in acoustic surveys of Norwegian spring-spawning herring wintering in the Ofotfjorden–Tysfjorden system.

allocation are stored in retrievable form, as in a database, with such identification as location and time of data collection.

Based on knowledge of fish species and size, in addition to characteristics of the transducer and echo-sounder system, the acoustic measures of fish density may be converted to biological measures, e.g. number of fish per unit volume or number of fish per unit area as projected onto the surface. In the case of the widely used technique of echo integration, this is accomplished by dividing the area backscattering coefficient by the mean backscattering cross-section (Foote, 1993). Other operations may be necessary with this technique if, for example, the sampling volume is different from the nominal value (Foote, 1991) or acoustic extinction is significant (Foote, 1990).

Once the fish density has been determined along the line transects spanning the survey area, the total abundance may be computed. This is

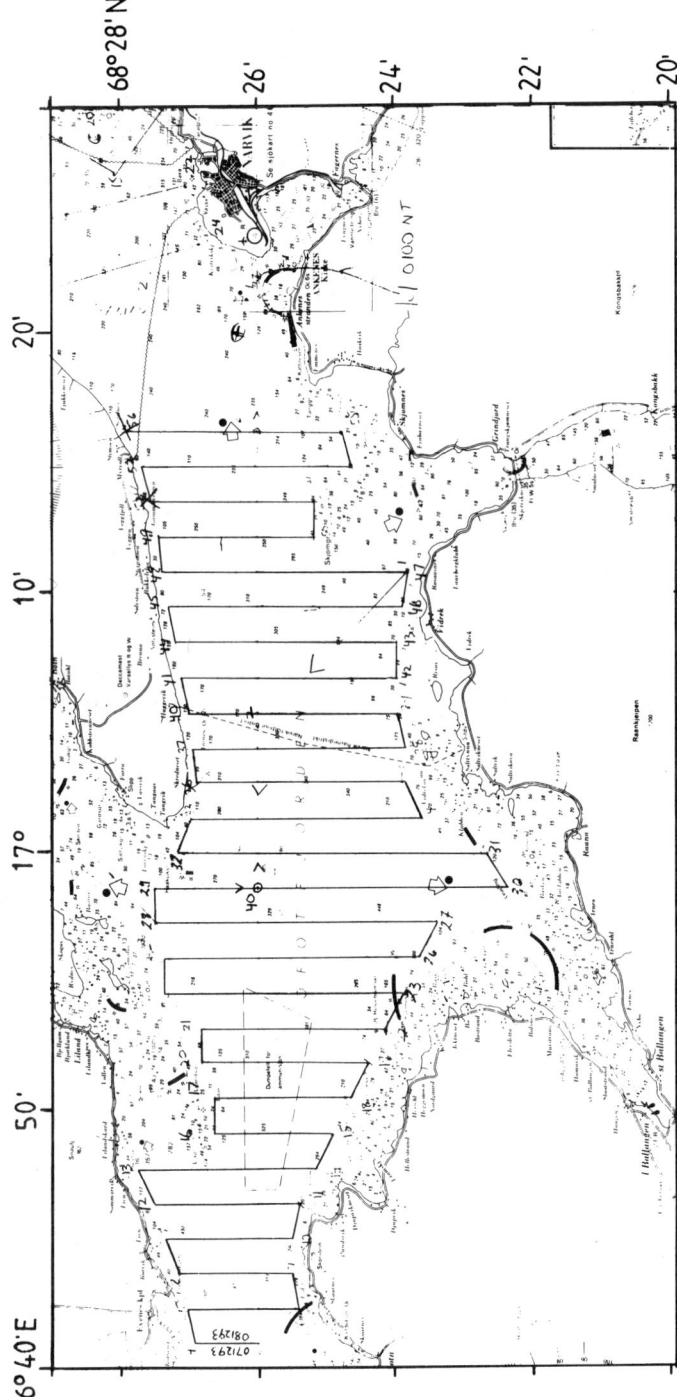

Fig. 4.2 Fine-scale survey design composed of equally spaced N–S parallel transects applied to the predominant herring concentration, in central Ofotfjorden, December 1993.

the integral of density over the entire area or volume. Since only partial information is generally available, gathered along the ship track, interpolation is generally necessary. This may however be accomplished implicitly, as by assuming that each measurement of density is equally representative of the mean as any other, as related to the particular transect or the entire area. The stock abundance is generally estimated from the density measurements by numerical integration. This number is often partitioned by age or size class, thus resulting in a series of abundance estimates that pertain to the several classes.

Examples of acoustic abundance surveys are described in Jakobsson (1983) for herring and in Traynor and Nelson (1985) for walleye pollock (*Theragra chalcogramma*). An illuminating exposition of the general technique is given in Gerlotto and Stéquert (1983). This is based on survey situations exemplifying high degrees of homogeneity or heterogeneity in spatial distribution. Johannesson and Mitson (1983) describe a wide range of survey types, while also presenting the basic methodology in heuristic fashion.

It may be evident that the described measurements, signal processing, data storage, retrieval, and analysis operations are computationally intensive. This may be clearer from a consideration of the quantities of data that are typically collected during a survey. These are described following reviews of two basic devices, the echo sounder and the echo integrator.

4.5.2 Echo sounder

In its simplest form, an echo sounder is a box of electronics that controls transmission of an underwater signal by a transducer and reception of echoes by the same device. The primary purpose of an echo sounder is to determine the range to interesting scatterers or targets, for example, the bottom or fish. It generally does this through transmission of a short signal, called a ping, conversion of the echo pressure fluctuations to an electrical signal, which may then be amplified and filtered, and display of the electrical echo signal, as on a paper chart or visual display unit. When the echo amplitude is represented by brightness or color and each time series of amplitudes is displayed versus time or range, and aligned with the preceding time series, the display is called an echogram. Examples are presented in Plate 2.

A scientific echo sounder performs the same function as an ordinary echo sounder but with provision of a calibrated output signal. This signal makes possible quantitative use of echo information, as in echo integration, described below.

Calibration may be effected in a variety of ways (Robinson, 1984; Urick, 1983; Sawada and Furusawa, 1993). The most widely used technique employs a so-called standard target consisting of a solid elastic sphere whose material and diameter depend on the transmit frequency. Detailed guidelines for calibrating an echo sounder are available in Foote *et al.* (1987).

The calibrated output signal generally but not exclusively incorporates a particular form of time-varied gain (TVG) or range compensation. In one form, the echo signal is amplified in the intensity domain by the function $r^2 10^{2\alpha r/10}$, where r is the range, $r = ct/2$, c is the speed of sound, t is the echo time relative to the start of signal transmission, and α is the absorption coefficient in decibels per unit distance. In the logarithmic domain, the TVG function is $20\log r + 2\alpha r$. This TVG function is useful for comparing the echo strength of layers of scatterers: the calibrated output signal from identical layers will, in a statistical sense, be independent of range.

A second common form of TVG is that of $r^4 10^{2\alpha r/10}$ in the intensity domain or $40\log r + 2\alpha r$ in the logarithmic domain. This TVG function is useful for comparing the echoes from isolated scatterers. The calibrated output signal from the same single-body target will be independent of range when this TVG function is applied.

Historically, TVG was applied by analog circuitry. When applied on a digital computer, the term is properly referred to as range compensation.

4.5.3 Echo integrator

An echo integrator is a device or set of computer instructions that integrates the intensity or square of the calibrated output signal with '$20\log r + 2\alpha r$' range compensation or like quantity. Briefly, this is done to derive a measure proportional to the energy in the echo signal, which under typical conditions and in the mean of a large number of echoes, is proportional to the sum of the echo energy contributions from the individual scatterers, hence to fish number density (MacLennan, 1990).

If the energy-proportional quantity is the volume backscattering coefficient s_v, then the result of echo integration over the depth range (z_1, z_2) is the area backscattering coefficient,

$$s_a = \int_{z_1}^{z_2} s_v(z)\, dz$$

where the echo sounder is assumed to be used in its ordinary downward-looking mode. Since s_v represents the cumulative backscattering cross-section of all scatterers in the acoustic sampling volume bounded by the described depth range, s_a has the dimensions of scattering area per unit sampling area. In fisheries research, it is convenient to express the area backscattering coefficient through a proportional quantity (Knudsen, 1990),

$$s_A = 4\pi 1852^2 s_a$$

In conventional SI units, s_a has the dimensions of square meters of back-scattering cross-section per square meter, hence s_A has the dimensions of square meters of backscattering cross-section per square nautical mile (NM), using the nominal conversion factor $1\,\text{NM} = 1852\,\text{m}$.

The smallest significant value of s_A that is typically encountered with commercially important fish is $1\,\text{m}^2\,\text{NM}^{-2}$. Maximum values achieved for very dense and extended fish schools may exceed $10^6\,\text{m}^2\,\text{NM}^{-2}$. Values for flat sea bottoms, if integrated, are often larger.

For applications to biological scatterers, s_A is expressed in terms of the area density ρ_A and characteristic mean backscattering cross-section $\bar{\sigma}$ of the scatterers through the fundamental equation of echo integration,

$$s_A = \rho_A \bar{\sigma}$$

The backscattering cross-section σ is related to target strength TS through the definition

$$\text{TS} = 10\log \frac{\sigma}{4\pi r_o^2}$$

where r_o is a reference distance, typically assumed to be $1\,\text{m}$. A similar equation applies to $\bar{\sigma}$. Given some relationship between TS and mean or root-mean-square fish length l of an essentially monodisperse aggregation of fish, the TS–l equation is solved for TS, which is reduced to $\bar{\sigma}$, which divides s_A to yield a value for ρ_A.

In addition to performing the described integration, echo integrators often also average corresponding integrator values over particular intervals of sailed distance, as selected or specified by the operator. Use of data from the ship log is thus implied, for determining the number of pings to be averaged. When integration includes averaging, compensation for changing vessel speed is usually automatic.

4.5.4 Data quantities

The number of acoustic data that are typically collected and stored during acoustic surveys spans many orders of magnitude. Two current examples are derived from the annual winter survey of Norwegian spring-spawning herring in the Ofotfjord–Tysfjord system. In each example, single-ping data are represented by values of the volume backscattering strength $S_v = 10\log s_v$ with 1 m or 2 m depth resolution over a depth range of 500 m or 1000 m, respectively. An additional 150 samples are collected in the so-called bottom channel. Each value requires two bytes for its digital representation. Thus the quantity of data associated with a single echo time series is 1.3 kbyte. Echo data are typically collected simultaneously on 1–4 transducers. At a pulse repetition frequency of 50 pings per minute, the nominal data rate is 4–16 Mbyte per hour.

In order to survey the mentioned fjord system, indicated in Figure 4.1, with total area of about 200 square nautical miles, abbreviated NM^2 here, about 40 h of sailing time is required. When all four transducers are operated simultaneously, the total quantity of collected data is 640 Mbyte.

Because of the complex shape of the combined Ofotfjord–Tysfjord system and behavioral patterns of the herring that cause the spatial distribution to be irregular, the fjord is often stratified for surveying purposes. For a particular stratum in Hulløysund, the herring may be concentrated in an area of about 9 NM^2. In a survey lasting 2 h and involving only a single transducer, the total quantity of collected data is 8 Mbyte.

Clearly there are fish stocks that occupy lesser or much greater areas, say from a fraction of 1 NM^2 to 100 000 NM^2, requiring survey times from 1 h to 2–3 months. Total data quantities are thus in the nominal range 1 Mbyte– 10 Gbyte. Use of a digital computer to help analyze these is essential.

4.5.5 Scientific echo sounder systems

A total of five scientific echo sounder systems are reviewed. The manufacturers are BioSonics, Inc., Hydroacoustic Technology, Inc., Kaijo Company, MICREL S.A., and SIMRAD Subsea A/S. Because of the present orientation to large-scale surveys of marine fish stocks, the SIMRAD instrument is reviewed first and in greater detail than the other, newer systems. Clearly, however, each has exploited state-of-the-art technology and has a very considerable potential for improving the measurement of fish density, hence estimates of fish stock abundance.

SIMRAD EK500 scientific echo sounder

The scientific echo-sounder system that is most widely used by major marine research institutions is the SIMRAD EK500 scientific echo sounder (Bodholt *et al.*, 1989). This system serves as an echo sounder, echo integrator, and, with split-beam transducer, target strength analyzer. Its characteristics and operation are briefly summarized.

Multiple-frequency operation
The EK500 may drive one, two, or three transducers simultaneously, with parallel processing of the respective received signals. Transceivers are available for the following standard frequencies: 12, 18, 27, 38, 49, 120, and 200 kHz. Transducers may be single-beam at any frequency or split-beam at 18, 38, and 120 kHz.

Dynamic range
Use of logarithmic amplifiers in the receiver has achieved a dynamic range of 160 dB, or a ratio of 10^{16}:1 with respect to intensity or s_v, and 10^8:1 with respect to amplitude. This dynamic range spans signal strengths from very low ambient noise to echoes from a hard, flat sea bottom.

Range compensation
Traditional time-varied gain is supplanted by range compensation that is applied in a built-in microprocessor. The sound speed profile, as determined by STD-sonde, for example, may be arbitrarily specified by the operator.

Bottom detection
This function was first implemented in the SIMRAD EA500, a hydrographic echo sounder designed for bottom mapping. It is based on a so-called point system, whereby points are awarded to candidate echoes on the basis of, for example, echo strength, continuity of echo strength, continuity of depth, operator-selected criteria such as depth range. The candidate with the greatest number of points is defined as the bottom.

Echo integration
The master processor performs echo integration on each of the transducer signals. Absolute values of volume backscattering strength S_v or area backscattering coefficient s_A are computed internally for predefined layers and sublayers. Layers may be defined with respect to the surface, both with and without requirement of bottom detection, or with respect to the bottom.

Target strength analysis
When the echo-sounder system is used with a split-beam transducer, the target strength of resolved single-fish echoes can be directly measured (Ehrenberg, 1979). This is done in the master processor.

Operator control
Operating parameters are set through a menu system, which is arranged in a hierarchy. Control is thereby exerted over displays, recorders, transceivers, bottom detection criteria, integration layers, target strength detection criteria, among others.

Data communications
Echo-sounder data are sent to a color monitor or recorders for display as echograms. Results of echo integration and target strength analysis may be displayed in tabular form on the printed echogram, if so directed through the menu system. The target strength data may also be displayed through histograms on the color monitor. Results of echo integration and target

strength are broadcast through serial outputs. The digitized echo time-series with longitudinal and transverse angles and beam-pattern-compensated amplitudes are available on a parallel output port. The various data are also broadcast over a local area network (LAN) through an Ethernet port with Transport Control Protocol/Internet Protocol (TCP/IP).

BioSonics DT4000 scientific digital acoustics system

This is the latest model in a series of scientific echo-sounder systems manufactured by BioSonics that aim at numerous small-scale survey applications, which do not require the robustness or power of instruments used to survey large sea areas. Digitization has been advanced very near to the transducer, thus establishing the enviable potential for minimum noise processing. The system, exclusive of printer and mass storage device, is eminently portable, fitting in a briefcase-sized unit. The system combines the functions of echo sounding, echo integration, and, with single-beam transducer, target strength analysis by deconvolution. Transducer frequencies from 4 to 1000 kHz can be accommodated. Both ordinary square-wave-modulated and chirp signals are available, the second type achieving a 10 dB enhancement of the signal-to-noise ratio.

HTI Split-beam system

Hydroacoustic Technology, Inc., has designed a scientific echo-sounder system on the basis of modules, which together embrace the functions of echo sounding, echo integration, and, with split-beam transducer, target tracking and target strength analysis. The separate parts of the system are the Model 240 split-beam digital echo sounder, Model 402 digital chart recorder, Model 440 digital tape interface, Model 464 digital multiplexer, Model 340 digital echo processor, and Model 540 split-beam transducer and cables. An IBM PC/AT compatible computer is an additional component necessary for analysis of echo integrator and single-target data. Two transducers can be simultaneously sampled. The standard transducer frequency is 200 kHz, but alternatives are 38, 120, 420, and 720 kHz. The total dynamic range is 140 dB. Notably, the Model 240 determines single-target angular position by quadrature demodulation. This method is demonstrably superior to that of zero-crossing detection, which is employed in the SIMRAD EK500 echo sounder, especially at frequencies above about 100 kHz. Recent advances in the HTI system include addition of FM-slide or chirp signal, with promise of raising the signal-to-noise ratio by 15 dB, in the Model 240 DES, and eight-transducer operating capability in the Model 242 DES with internal multiplexer. With external multiplexer, as many as 16 transducers can be operated.

Kaijo versatile echo sounding system (VESS)

This third-generation system combines the functions of echo sounding, echo counting, echo integration, and, with split-beam transducer, target strength analysis. The operating frequencies are 38 and 70 kHz, with transducers operating simultaneously in both dual-beam and split-beam modes. A number of novel operations are performed, for example, that of echo integration with each of the dual beams, both narrow and wide, so as to assess the effect of avoidance reaction on acoustic measurements of fish density. Results of such operations are displayed. In the particular example, the difference in levels of backscattering strength between the two beams is displayed. Documentation of this system is provided partially through a description of the second-generation system (Furusawa *et al.*, 1993) and through an explication of the principles underlying the design of quantitative echo sounders (Furusawa, 1991). The second-generation system differs from the third-generation system mainly by operating only one transducer, at 38 kHz, in dual-beam mode only.

MICREL OSSIAN 1500 echo sounder

This is a combined echo sounder and echo integrator system produced by MICREL S.A. It works at high power, up to 5 kW, for deep-bottom detection. It may drive one or two transducers at the same time. Operating frequencies of single-beam transducers are 12, 28, 30, 38, 50, 88, 120, and 200 kHz; those of dual-beam transducers are 38 and 120 kHz. Both $(20\log r + 2\alpha r)$ and $(40\log r + 2\alpha r)$ TVG functions are available. In addition, two independent, adjustable gains are available, for application to the water column and bottom, respectively. Fish schools are recognized automatically, with characterization by three sets of parameters, based on image shape, bathymetry, and echo energy (Weill *et al.*, 1993). Echoes from detected fish schools are automatically integrated. The upper layer of the sea bottom is also automatically classified. Data communication is achieved by alternative means, including LAN. IFREMER and ORSTOM, among other institutions, both use the OSSIAN Model 1500.

4.5.6 Postprocessing system

Until about 1990, nearly every major marine research institute that surveys fish stocks acoustically used its own digital computer system for the postprocessing of acoustic and related survey data. This generally involved software that was developed for a particular configuration of computer hardware, both in terms of processor types and manufacture. Changing computer generally meant changing software too. The software was, moreover, often written in a low-level language and assembled in a mono-lithic program, requiring sequential execution of the entire program, if with options or branch points, no matter what the scope of the particular exercise.

To remedy this situation and to achieve a much greater degree of flex-ibility in the postprocessing of survey data, development of a new system was begun at the Institute of Marine Research, Bergen, in 1988. This system, called the Bergen Echo Integrator (BEI) (Foote *et al.*, 1991), has also spawned a commercial version, the SIMRAD BI500, with world-wide distribution and use *per* 1994. As its design principles are quite general and since it may serve as the model for other new systems, the basic develop-ment of BEI is described. Its context is indicated in Figure 4.3, which sketches the layout of a data network on board a research vessel.

User requirements and basic needs

At an early stage in planning the system, users imposed the following requirements:

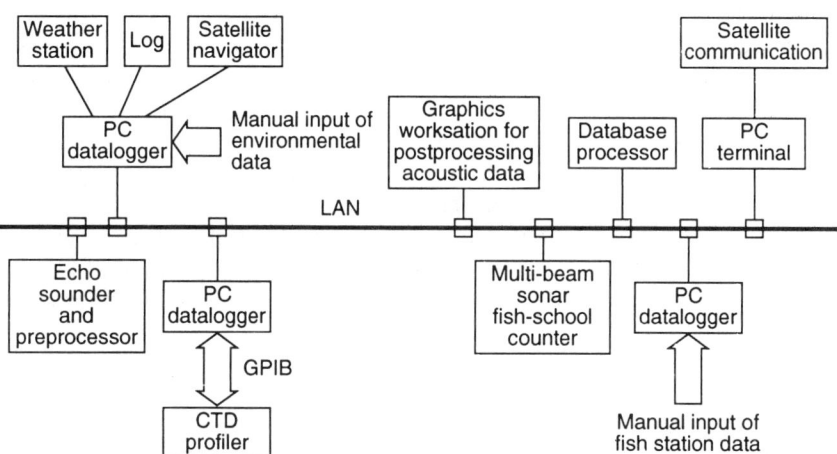

Fig. 4.3 Layout of a research vessel data network (reproduced with permission, H.P. Knudsen).

(a) adequate capacity to process, display, and manipulate data from an echo sounder preprocessor;
(b) machine independence;
(c) high-level language for programming;
(d) modular construction;
(e) database;
(f) Ethernet local area network (LAN);
(g) full documentation;
(h) user friendliness.

Several basic needs are thus identified. First, the system should be user-specified, i.e. the user should make all important decisions. Secondly, the system should be easy to use; if it is not, it will not be used. Thirdly, the system should be capable of expansion, hence the architecture should be extendable and scalable. Finally, full documentation is indeed essential, so as to facilitate future expansion.

Computing environment

Hardware

The system is designed to serve as a data logger for the EK500 echo sounder. The daily data rate is consequently 100–400 Mbyte. The system is dimensioned to store data continuously for 24 h before cleanup procedures must be started. Data are displayed in echograms consisting of 650 S_v values per ping – 500 values in the water column and 150 values about the detected bottom. An echogram should be displayed with transformations, such as changing the noise threshold, and computations, such as recomputing area backscattering coefficients, within 10 s. The raster graphics should render the original echogram without information loss. The graphics system should include a frame buffer with 256 colors from a modifiable color palette. The estimated speed of the system is thus 10 million instructions per second (MIPS), 2 million floating point operations per second (MFLOPS), and input/output (I/O) bandwidth of 2–3 Mbyte per second. Reduced instruction set computer (RISC) architecture was thus required at the outset of the project. The hardware technology that fulfilled the requirements in 1988 is that of the workstation class of machines.

Software

The requirement of an open architecture is satisfied by employing international, nonproprietary standards. The operating system UNIX (Christian and Richter, 1994) was chosen for portability, multitasking capacity, virtual memory, nonsegmented memory, and networking. The programming language C (Kernighan and Ritchie, 1988) was chosen for its support of both structured programming and efficient coding, with powerful interface to UNIX. A relational database, INGRES (Date, 1987a), was chosen, with standard structured-query language (SQL) (Date, 1987b). A user-friendly interface was ensured by choosing that graphical user interface (GUI) which was both nonproprietary and portable, namely the X Window system (Young, 1994).

External communications

Choice of a local area network (LAN) avoided the limitations inherent in the traditional monolithic system with peripherals attached directly to a central processing unit (CPU). The most widely used LAN interface was chosen, namely Ethernet and Transport Control Protocol/Internet Protocol (TCP/IP). A subset of TCP/IP was selected for data acquisition, namely User Datagram Protocol/Internet Protocol (UDP/IP). The advantage of this protocol is that data can be sent to the receiver even during times of temporary overflow, albeit with data loss, which is noncritical apropos of the quantities involved.

System design

Data flow

High-volume data, such as the raw acoustic data from the echo-sounder preprocessor, are stored directly in files, while information on their access is stored in the database. Postprocessed data and other low-volume data, such as navigation data and salinity-temperature-depth (STD) data, are stored directly in the database. The system displays and processes data that are accessed from both database and files. Upon completion of the postprocessing operations, so-called report generators summarize the results.

Configuration window

Key data on the cruise are specified through the configuration window. These typically include nation, vessel name, and cruise number to associate data in the database, but also purpose, plan, sea area, and cruise personnel. Channel depths for echo integration are included, as is the basic interval of sailed distance over which echo integration is performed. Typical layer depths for echo integration are 50 m for pelagic fish and 1, 2, 5, and 10 m for bottom fish as measured from the detected bottom. Typical intervals of sailed distance are 0.1, 0.5, 1, and 5 nautical miles.

Windows for data postprocessing

(a) *Survey grid window* – this displays the location of cruise data stations; for example, stations where echo sounder, trawl, oceanographic, and other data are collected.

(b) *Echogram window* – this consists of the following four subwindows. The main echogram subwindow displays the echogram in a form suitable for postprocessing by the operator. The expanded bottom channel subwindow displays the echogram expanded about the detected bottom, with possibility of operator correction of this by redrawing the bottom line. The color map subwindow specifies the color palette. This may be defined by the operator, but is typically selected from among three choices: gray scale for observing shape, red–blue color scale for gauging signal strength, and dark red–light blue color scale as a compromise of the first two listed scales, combining the features of shape and signal discrimination. The zoom subwindow allows expansion of arbitrary, operator-selected parts of the main echogram subwindow or expanded bottom channel subwindow.

(c) *Interpretation window* – results of echo integration according to operator delineation of the echogram are stored in the database by scatterer type. This window makes available names of scatterer classes, from database or by operator definition, for assignment of integration values.

(d) *Target strength window* – target strength values of resolved single-target echoes may be selected, compiled, and displayed in a histogram for operator-delimited regions of the echogram.

(e) *File selection window* – acoustic data may be selected through this window, which might be useful when navigation data are lacking.

(f) *Fish station window* – biological data derived by tracking or other means may be summarized and displayed in the form of length distribution and number of measured specimens for each selected species that is represented in the catch data. The selection may be performed in the survey grid window.

(g) *STD window* – STD profiles are displayed, as designated in the survey grid window.

System implementation

Integration techniques

It should be possible to activate a subset of the system without activating the whole. Since the system is represented by a set of processes, each of which executes its own task independently of the others, communication among the processes is essential. This is achieved through the communication mechanisms available in the window system itself. Procèsses announce their entrance to and departure from other window-based processes.

Performance issues

(a) If mechanisms for database I/O had sufficient bandwidth, all data could be stored in the database. Since this is most definitely not the case, data are separated into two categories, according to the volume of data. As already mentioned, high-volume data are stored in files, while low-volume data are stored in the database. Refined data, such as results of echo integration, are stored in the database.

(b) CPU-intensive parts of the system are optimized by precalculating mathematical expressions and storing the results in index tables. An example is afforded by the operation of echo integration: an S_v value, which is expressed as a logarithm to base two, is used as an index to a vector of precalculated s_A values.

Summary of special features

Five special features of the Bergen echo integrator are enumerated:

(a) The user may delineate arbitrarily shaped regions of the echogram for echo integration, including non-constant-depth intervals, by means of interactive graphics.

(b) Results of echo integration may be stored in the database with different degrees of resolution. In terms of sailed distance, these are 0.1, 0.5, 1, and 5 nautical miles. In terms of depth, these are nearly arbitrary, but typically lie in the range from 0.1 to 500 m.

(c) Errors made in preprocessing may be corrected. Examples include re-definition of the detected bottom and changing the noise threshold.

(d) Different color maps may be used to aid the extraction of information on shape and signal strength from the echogram.

(e) Interconnections of graphical user interfaces, database, and data files are designed to optimize data flow and the exercise of system functionality.

Some current applications

The described postprocessing system is being used in routine survey work as well as in a number of special applications. These include collection of data for analysis of structure, through the variogram, and exploitation of this in computing the estimation variance associated with abundance estimates, studies of target strength, and studies of fish behavior, among others.

4.5.7 Sonar

Sonars have been used in abundance estimation of fish stocks through counting fish schools (Hewitt *et al.*, 1976). Based on knowledge or assumption of individual school size, biomass can be assessed. As in the case of conventional echo-sounder measurements of fish, density is measured along line transects and integrated over the survey region in order to estimate total abundance.

Rather recent work (Misund *et al.*, 1992; Misund, 1993) aims to exploit the two-dimensional imaging capability of sonar. This uses the SIMRAD SA950 sonar, with 95 kHz operating frequency, when operated in a sector-scanning mode with 45° horizontal sector spanned by 32 beams, each with 1.7° horizontal beamwidth and 10° vertical beamwidth. A variety of pulsed sinusoidal and FM signals are available; an FM signal is generally used for its enhanced definition of target position. Through extensive field trials in which detected fish schools have been subsequently caught by seining, horizontal fish school area has been quantitatively related to school biomass. The relationship is presently imprecise, but already the technique is reckoned to significantly improve abundance estimates of certain schooling fish, which may avoid vessels with vertically oriented echo sounders or which may at times occur near the surface, where they are inaccessible to an echo sounder beam.

4.6 INTEGRATED SURVEYS

Study of the literature can only confirm the individual investigator's experiences of the difficulty of performing a fish stock abundance survey. Given the diversity of measurements ordinarily performed on such surveys, the attraction of combining essentially independent methods is powerful. Here, two examples are outlined.

4.6.1 Combined acoustic and egg-production survey

Acoustic and egg-production surveys have been combined in order to assess the spawning stock biomass of Cape anchovy (*Engraulis capensis*; Hampton *et al.*, 1990). During the same survey, both methods were essentially independently applied. The echo integration method was applied over the targeted region of occurrence, the South African continental shelf. Eggs were counted at regular intervals along the acoustic transects, and the proportion of spawners was estimated by pelagic trawl. Estimates of spawning biomass were derived by the mean daily-egg-production method. Agreement of the two estimates for each of the reported years, 1985 and 1986, is excellent. Given the intrinsic difficulties associated with the execution of most surveys, the agreement adds much confidence to the estimates at very little additional cost.

4.6.2 Integrated acoustic and trawl survey

Physical capture is an essential part of most acoustic surveys. It is generally necessary for both species identification and determination of size or age composition of the acoustically observed fish. The integral nature of trawling to acoustic surveying is emphasized in Thorne (1983) and Traynor and Nelson (1985).

In many studies, however, bottom fish that might otherwise be acoustically measured when off the bottom are simply not accessible, because of the so-called bottom dead zone (Mitson, 1983). It is precisely this realization, coupled with a wider appreciation of the problems of sampling by trawl, that may lead to an integration of acoustic and trawl surveys, with joint performance of the two survey types (Godø and Wespestad, 1993).

4.7 AUXILIARY INSTRUMENTS

Many instruments may be construed as auxiliary in the context of research surveys. Here, just three examples are considered, for their widespread use or recognized potential in the measurement of major oceanographic features, which evidently influence the occurrence of fish in the water column.

4.7.1 Salinity-temperature-depth sonde

The salinity-temperature-depth (STD) sonde is a traditional profiling instrument that measures the temperature and salinity fields in the water column. This is usually achieved through direct measurements of conductivity, temperature, and pressure. Based on these, the sound speed profile can be computed (Mackenzie, 1981) and used in the range compensation function applied to echo-sounder signals. The STD measurements are also used when interpreting echograms, because of the recognized association of fish and oceanographic features. Examples of displays of information resulting from hydrographic surveys in the Norwegian Sea are given in Blindheim (1990).

Data quantities can be formidable. For the Neil Brown III CTD System, for example, 150 bytes are used to represent the data triplet of pressure, conductivity, and temperature. At a sampling frequency of $31\,Hz$ and typical lowering speed of $1\,m\,s^{-1}$, the data rate is $4.65\,kbyte\,s^{-1}$. For a depth range of $500\,m$, the total quantity of data is $2.3\,Mbyte$, expressed in binary format. Use of recursive and low-pass filtering affects averaging, with loss of resolution and reduction in data quantity. Nominal accuracies are given in the cited article (Blindheim, 1990).

4.7.2 Acoustic Doppler current profiler

The acoustic Doppler current profiler (ADCP) measures current velocity throughout the water column, conditional on the presence of passive scatterers. Such information is· potentially important in interpreting echosounder data for the same reason that the STD profile is important: fish occurrence is often closely linked to the local oceanographic conditions. The same information is important for determining advection in egg and larval surveys.

The device works by detecting the movement of relatively large or small but concentrated scatterers at a succession of depths throughout the water column. Accomplishing signal detection in order to employ the Doppler principle to measure velocity also suggests the potential for using an ADCP for the direct measurement of biological scatterers. This has been done with modification of the instrument (Flagg and Smith, 1989) and without modification (Heywood *et al.*, 1991). The device is thus used in the respective cases as an echo sounder with or without a calibrated output signal.

If capable of being calibrated with respect to signal strength, then given sufficient computing capacity, an ADCP may serve as the primary instrument in a quantitative acoustic survey. Already, information from an ADCP has been combined with coincident hydrographic information from an undulatory sonde with 500 m vertical range (Roe and Griffiths, 1993). Three commercial ADCPs are described in Cochrane *et al.* (1988). An ADCP manufactured by RD Instruments, with operating frequency of 307.2 kHz, is described in Flagg and Smith (1989).

Data rates associated with an ADCP are comparable to those from a scientific echo sounder of similar frequency, assuming that only the equivalent sum-beam signal is recorded. If each of the individual beam output signals were to be recorded, then the raw data rate would be comparable to that from a split-beam transducer before signal processing, i.e. four times the ordinary sum-beam rate.

4.7.3 Acoustic bottom-classifier system

Bottom fish show preference for bottom type (Scott, 1982; Orlowski, 1989; Walsh, 1992). According to the theory developed by Orlowski (1984), the ratio of energy contained in the bottom–surface–bottom echo with energy contained in the single bottom-bounce echo is related to bottom hardness and roughness. In fact, a commercial device, called RoxAnn, is used by both fishermen and researchers (Burns *et al.*, 1989), but with varying success.

The device is essentially an echo sounder with special signal processor.

The possibility of extending performance in the future by processing alone, as by application of neural networks on the basis of control data, is evident.

4.8 POTENTIAL SURVEY APPLICATIONS OF NEW OR UNIQUE ACOUSTIC INSTRUMENTS OR TECHNIQUES

A number of new or unique acoustic instruments or techniques could be applied in research surveys aimed at determining fish stock abundance. Three are considered here.

4.8.1 Multiple-echo-sounder beams

Ordinary echo integration surveys generally employ only a single, vertically oriented, directional beam. The sampling volume is thus relatively small. Addition of a second, similar beam, oriented to the side of the first, would double the sampling volume. It would also facilitate detection of possible fish reactions to the passage of the survey vessel.

In the more general case of a sector-scanning sonar (Mitson and Cook, 1971) or high-resolution fan-beam echo sounder (Iida and Suzuki, 1987; Aoki *et al.*, 1991), addition of suitable electronic circuitry and computer processing can allow echo integration to be performed on each individual beam. The dual advantages are a potentially very substantial increase in sampling volume and possibility of imaging the three-dimensional structure of a fish distribution, including quantifying possible fish reactions to passage of the research vessel (Foote, 1979). The mentioned necessary addition of computer processing power could be met by a machine with parallel processing.

4.8.2 Multifrequency acoustic profiling system

The multifrequency acoustic profiling system (MAPS) (Holliday *et al.*, 1989) is a sonde with 21 planar, resonant-frequency transducers, with frequencies logarithmically spaced over the range from 100 kHz to 10 MHz. This unique system has been used in a number of special investigations, generally executed along single transects, to study the distribution of zooplankton in relation to oceanographic features. The zooplankton distribution has been characterized by animal size, number density, and location, both in depth and in horizontal distance along line transects.

The surveying potential of such an instrument, as for determining the abundance of zooplankton or larval fish or other 0-group fish over a sea area, has been recognized within the Global Ocean Ecosystem Dynamics (GLOBEC) Program (US GLOBEC, 1991; International GLOBEC, 1993). Both smaller- and larger-scale designs of acoustic sondes have been proposed. As in the case of echo integration surveys with multiple-echo-sounder beams, demands on computer processing can be formidable. Parallel processing is clearly required. Display of information is similarly computationally intensive. Integration of the density distribution over an area can be accomplished by standard techniques, including use of geostatistics. The amount of computation may be considerable in view of the potential number of scatterer classes, including both species and size and possibly depth too.

Data rates associated with MAPS may be inferred from its description in Holliday *et al.* (1989), assuming a specific sampling rate. Given a nominal pulse duration of $50\,\mu s$, hence with nominal range resolution of $37.5\,mm$, and sampled range interval $1–2\,m$ from the sonde, a sampling period of $50\,\mu s$ will yield 27 samples per sampled range interval per frequency. At a pulse repetition rate of 21 frequencies per $21/30\,s$, the basic data rate is 540 samples per second. If each sample were to be represented by 2 bytes, then the basic data rate would be $1\,kbyte\,s^{-1}$, which is comparable to that of a scientific echo sounder operating at $38\,kHz$ with pulse repetition frequency of 50 pulses per minute.

4.8.3 Multiple-frequency echo integration

Frequency diversity in measurements of marine biological scatterers continues to find new applications since its description in Holliday (1980), and early application in an instrument (Holliday and Pieper, 1980). One is in echo integration carried out essentially continuously along line transects. The application is being developed at the Institute of Marine Research, Bergen, with simultaneous echo integration at each of four frequencies, 18, 38, 120, and $200\,kHz$. At least three advantages can be identified:

(a) redundancy, as in guarding against transducer failure;
(b) separation of scatterer classes in interpreting echo sounder data, because of frequency dependent differences in scattering strength;
(c) possible inference of behavior, hence target strength, because of the frequency dependence of area backscattering coefficient.

Such a technique could enable the backscattering cross-section, or target strength, of fish to be determined *in situ* without regard to number density.

ACKNOWLEDGEMENTS

E. Moksness, who proposed the topic, is thanked for advice during preparation of the manuscript. Other institute colleagues are also thanked, especially A. Aglen, J. Blindheim, H.P. Knudsen, E. Ona, I Røttingen, S.

Sunby, and O.J. Østvedt. M. Aksland is thanked for comments on the mark–recapture method.

REFERENCES

Aasen, O. (1958) Estimation of the stock strength of the Norwegian herring. *Journal du Conseil International pour l'Exploration de la Mer*, **24**, 95–110.

Ådlandsvik, B. and Sundby, S. (1994) Modelling the transport of cod larvae from the Lofoten area. *ICES Marine Science Symposia*, **198**, 379–92.

Anthony, V.C. and Waring, G. (1980) The assessment and management of the Georges Bank herring fishery. *Rapports et Procès-Verbaux des Réunions. ICES*, **177**, 72–111.

Aoki, Y., Sato, T., Zeng, P. and Iida, K. (1991) Three-dimensional display technique for fish-finder with fan-shaped multiple beams, in *Acoustical Imaging*, (eds H. Lee and G. Wade), Plenum Press, New York, Vol. 18, pp. 491–9.

Astthorsson, O.S., Gislason, A. and Gudmundsdottir, A. (1994) Distribution, abundance, and length of pelagic juvenile cod in Icelandic waters in relation to environmental conditions. *ICES Marine Science Symposia*, **198**, 529–41.

Blindheim, J. (1990) Arctic intermediate water in the Norwegian Sea. *Deep-Sea Research*, **37**, 1475–89.

Bodholt, H., Nes, H. and Solli, H. (1989) A new echo-sounder system. *Proceedings of the Institute of Acoustics*, **11**(3), 123–30.

Burns, D.R., Queen, C.B., Sisk, H., Mullarkey, W. and Chivers, R.C. (1989) Rapid and convenient acoustic sea-bed discrimination. *Proceedings of the Institute of Acoustics*, **11**(3), 169–78.

Burd, A.C. and Holford, B.H. (1971) The decline in the abundance of Downs herring larvae. *Rapports et Procès-Verbaux des Réunions. ICES*, **160**, 99–100.

Christian, K. and Richter, S. (1994) *The UNIX Operating System*, 3rd edn, Wiley, New York.

Cochrane, N.A., Whitman, J.W.E. and Belliveau, D. (1988) Doppler current profilers. *Canadian Technical Report of Fisheries and Aquatic Sciences*, (1641), 89–92.

Date, C.J. (1987a) *A Guide to INGRES*, Addison-Wesley, Reading, Massachusetts.

Date, C.J. (1987b) *A Guide to the SQL Standard*, Addison-Wesley, Reading, Massachusetts.

Dragesund, O. and Haraldsvik, S. (1968) Norwegian tagging experiments in the north-eastern North Sea and Skagerak, 1964 and 1965. *Fiskeridirektorates Skrifter Serie Havundersøkelser*, **14**, 98–120.

Dragesund, O. and Jakobsson, J. (1963) Stock strengths and rates of mortality of the Norwegian spring spawners as indicated by tagging experiments in Icelandic waters. *Rapports et Procès-Verbaux des Réunions. ICES*, **154**, 83–90.

Ehrenberg, J.E. (1979) A comparative analysis of *in situ* methods for directly measuring the acoustic target strength of individual fish. *IEEE Journal of Ocean Engineering*, **OE-4**, 141–52.

Ellertsen, B., Fossum, P., Solemdal, P. and Sundby, S. (1989) Relations between temperature and survival of eggs and first feeding larvae of the Arcto-Norwegian cod (*Gadus morhua* L.). *Rapports et Procès-Verbaux des Réunions. ICES*, **191**, 209–19.

Engås, A. and Godø, O.R. (1986) Influence of trawl geometry and vertical distribution of fish on sampling with bottom trawl. *Journal of Northwest Atlantic Fishery Science*, **7**, 35–42.

Engås, A. and Godø, O.R. (1989a) The effect of different sweep lengths on the length composition of bottom-sampling trawl catches. *Journal du Conseil International pour l'Exploration de la Mer*, **45**, 263–8.

Engås, A. and Godø, O.R. (1989b) Escape of fish under the fishing line of a Norwegian sampling trawl and its influence on survey results. *Journal du Conseil International pour l'Exploration de la Mer*, **45**, 269–76.

Flagg, C.N. and Smith, S.L. (1989) On the use of the acoustic Doppler current profiler to measure zooplankton abundance. *Deep-Sea Research*, **36**, 455–74.

Foote, K.G. (1979) Biasing of fish abundance estimates derived from use of the sector scanning sonar in the vertical plane, *Proceedings of Institute of Acoustics Conference – Progress in Sector Scanning Sonar*, Dec 18–19, Lowestoft, Institute of Acoustics, Edinburgh, pp. 44–52.

Foote, K.G. (1990) Correcting acoustic measurements of scatterer density for extinction. *Journal of the Acoustical Society of America*, **88**, 1543–6.

Foote, K.G. (1991) Acoustic sampling volume. *Journal of the Acoustical Society of America*, **90**, 959–64.

Foote, K.G. (1993) Application of acoustics in fisheries, with particular reference to signal processing, in *Acoustic Signal Processing for Ocean Exploration*, (eds J.M.F. Moura and I.M.G. Lourtie), North-Holland, Dordrecht, pp. 381–90.

Foote, K.G., Knudsen, H.P., Korneliussen, R.J., Nordbø, P.E. and Røang, K. (1991) Postprocessing system for echo sounder data. *Journal of the Acoustical Society of America*, **90**, 37–47.

Foote, K.G., Knudsen, H.P., Vestnes, G., MacLennan, D.N. and Simmonds, E.J. (1987) Calibration of acoustic instruments for fish density estimation: a practical guide. *ICES Cooperative Research Report*, (144), 1–69.

Fossum, P. and Moksness, E. (1993) A study of spring- and autumn-spawned herring (*Clupea harengus* L.) larvae in the Norwegian Coastal Current during spring 1990. *Fisheries Oceanography*, 2(2), 73–81.

Furusawa, M. (1991) Designing quantitative echo sounders. *Journal of the Acoustical Society of America*, **90**, 26–36.

Furusawa, M., Takao, Y., Sawada, K., Okubo, T. and Yamatani, K. (1993) Versatile echo sounding system using dual beam. *Nippon Suisan Gakkaishi*, **59**, 967–80.

Gerlotto, F. and Stéquert, B. (1983) Une méthode de simulation pour étudier la distribution des densités en poissons: application à deux cas réels. *FAO Fisheries Report*, (300), 278–92.

Godø, O.R. (1994) Factors affecting the reliability of groundfish abundance estimates from bottom trawl surveys, in *Marine Fish Behaviour in Capture and Abundance Estimation*, (eds A. Fernö and S. Olsen), Fishing News Books, Oxford, pp. 166–99.

Godø, O.R. and Engås, A. (1989) Swept area variation with depth and its influence on abundance indices of groundfish from trawl surveys. *Journal of Northwest Atlantic Fishery Science*, **9**, 133–9.

Godø, O.R. and Sunnanå, K. (1992) Size selection during trawl sampling of cod and haddock and its effect on abundance indices at age. *Fisheries Research*, **13**, 293–310.

Godø, O.R. and Walsh, S.J. (1992) Escapement of fish during bottom trawl sampling – implications for resource assessment. *Fisheries Research*, **13**, 281–92.

Godø, O.R. and Wespestad, V.G. (1993) Monitoring changes in abundance of gadoids with varying availability to trawl and acoustic surveys. *ICES Journal of Marine Science*, **50**, 39–51.

Gunderson, D.R. (1993) *Surveys of Fisheries Resources*, Wiley, New York.

Hampton, I., Armstrong, M.J., Jolly, G.M. and Shelton, P.A. (1990) Assessment of anchovy spawner biomass off South African through combined acoustic and egg-production surveys. *Rapports et Procès-Verbaux des Réunions. ICES*, **189**, 18–32.

Harden Jones, F.R., Margetts, A.R., Greer Walker, M. and Arnold, G.P. (1977) The efficiency of the Granton otter trawl determined by sector-scanning sonar and acoustic transponding tags. *Rapports et Procès-Verbaux des Réunions. ICES*, **170**, 45–51.

Haug, A. and Nakken, O. (1977) Echo abundance indices of 0-group fish in the Barents Sea, 1965–1972. *Rapports et Procès-Verbaux des Réunions. ICES*, **170**, 259–64.

Heessen, H.J.L. and Rijnsdorp, A.D. (1989) Investigations on egg production and mortality of cod (*Gadus morhua* L.) and plaice (*Pleuronectes platessa* L.) in the southern and eastern North Sea in 1987 and 1988. *Rapports et Procès-Verbaux des Réunions. ICES*, **191**, 15–20.

Hempel, I. and Hempel, G. (1971) An estimate of mortality in eggs of North Sea herring (*Clupea harengus* L.). *Rapports et Procès-Verbaux des Réunions. ICES*, **160**, 24–6.

Hempel, G. and Schnack, D. (1971) Larval abundance on spawning grounds of Banks and Downs herring. *Rapports et Procès-Verbaux des Réunions. ICES*, **160**, 94–8.

Hewitt, R.P., Smith, P.E. and Brown, J.C. (1976) Development and use of sonar mapping for pelagic stock assessments in the California current. *Fishery Bulletin*, **74**, 281–300.

Heywood, K.J., Scrope-Howe, S. and Barton, E.D. (1991) Estimation of zooplankton abundance from shipborne ADCP backscatter. *Deep-Sea Research*, **38**, 677–91.

Holliday, D.V. (1980) Use of acoustic frequency diversity for marine biological measurements, in *Advanced Concepts in Ocean Measurements for Marine Biology*, (eds F.P. Diemer, F.J. Vernberg and D.Z. Mirkes), University of South Carolina, Columbia, South Carolina, pp. 423–60.

Holliday, D.V. and Pieper, R.E. (1980) Volume scattering strengths and zooplankton distributions at acoustic frequencies between 0.5 and 3 MHz. *Journal of the Acoustical Society of America*, **67**, 135–46.

Holliday, D.V., Pieper, R.E. and Kleppel, G.S. (1989) Determination of zooplankton size and distribution with multifrequency acoustic technology. *Journal du Conseil International pour l'Exploration de la Mer*, **46**, 52–61.

Hylen, A., Korsbrekke, K., Nakken, O. and Ona, E. (1995) Comparison of the capture efficiency of 0-group fish in pelagic trawls, in *Precision and relevance of pre-recruit studies for fishery management related to fish stocks in the Barents Sea and adjacent waters*, Proceedings 6th Russian–Norwegian Symp., Bergen, 14–17 June 1994, (ed. A. Hylen), Institute of Marine Research, Bergen, Norway.

Iida, K. and Suzuki, T. (1987) High resolution fan-beam echo sounder. *Bulletin of Faculty of Fisheries, Hokkaido University*, **38**, 384–92.

International GLOBEC (1993) *Sampling and observational systems*, Report No. 3, GLOBEC International, Solomons, Maryland, USA.

Jakobsson, J. (1983) Echo surveying of the Icelandic summer spawning herring 1973–1982. *FAO Fisheries Report*, (300), 240–8.

Johannesson, K.A. and Mitson, R.B. (1983) Fisheries acoustics. A practical manual for aquatic biomass estimation. *FAO Fisheries Technical Paper*, (240), 1–249.

Jones, R. (1976) The use of marking data in fish population analysis. *FAO Fisheries Technical Paper*, (153), 1–42.

Kernighan, B.W. and Ritchie, D.M. (1988) *The C Programming Language*, 2nd edn, Prentice Hall, Englewood Cliffs, New Jersey.

Knudsen, H.P. (1990) The Bergen Echo Integrator: an introduction. *Journal du Conseil International pour l'Exploration de la Mer*, **47**, 167–74.

Lough, R.G., Bolz, G.R., Pennington, M. and Grosslein, M.D. (1985) Larval abundance and mortality of Atlantic herring (*Clupea harengus* L.) spawned in the Georges Bank and Nantucket Shoals areas, 1971–78 seasons, in relation to spawning stock size. *Symposia Northwest Atlantic Fishery Science*, **6**, 21–35.

Lough, R.G., Smith, W.G., Werner, F.E., Loder, J.W., Page, F.H., Hannah, C.G., Naimie, C.E., Perry, R.I., Sinclair, M. and Lynch, D.R. (1994) Influence of wind-driven advection on interannual variability in cod egg and larval distributions on Georges Bank: 1982 vs 1985. *ICES Marine Science Symposia*, **198**, 356–78.

Mackenzie, K.V. (1981) Nine-term equation for sound speed in the oceans. *Journal of the Acoustical Society of America*, **70**, 807–12.

MacLennan, D.N. (1990) Acoustical measurement of fish abundance. *Journal of the Acoustical Society of America*, **87**, 1–15.

Misund, O.A. (1993) Abundance estimation of fish schools based on a relationship between school area and school biomass. *Aquatic Living Resources*, **6**, 235–41.

Misund, O.A., Aglen, A., Beltestad, A.K. and Dalen, J. (1992) Relationships between the geometric dimensions and biomass of schools. *ICES Journal of Marine Science*, **49**, 305–15.

Mitson, R.B. (1983) Acoustic detection and estimation of fish near the sea-bed and surface. *FAO Fisheries Report*, (300), 27–34.

Mitson, R.B. and Cook, J.C. (1971) Shipboard installation and trials of an electronic sector-scanning sonar. *Radio and Electronics Engineer*, **41**, 339–50.

Munk, P. (1988) Catching large herring larvae: Gear applicability and larval distribution. *Journal du Conseil International pour l'Exploration de la Mer*, **45**, 97–104.

Munk, P. and Christensen, V. (1990) Larval growth and drift pattern and the separation of herring spawning groups in the North Sea. *Journal of Fish Biology*, **37**, 135–48.

Nakashima, B.S. and Winters, G.H. (1984) Selection of external tags for marking Atlantic herring (*Clupea harengus harengus*). *Canadian Journal of Fisheries and Aquatic Sciences*, **41**, 1341–8.

Nakken, O., Hylen, A. and Ona, E. (1995) Acoustic estimates of 0-group fish abundance in the Barents Sea and adjacent waters in 1992 and 1993, in *Precision and relevance of pre-recruit studies for fishery management related to fish stocks in the Barents Sea and adjacent waters*, Proc. 6th Russian–Norwegian Symp., Bergen, 14–17 June 1994, (ed. A. Hylen), Institute of Marine Research, Bergen, Norway.

Ona, E. and Godø, O.R. (1990) Fish reaction to trawling noise: the significance for trawl sampling. *Rapports et Procès-Verbaux des Réunions. ICES*, **189**, 159–66.

Orlowski, A. (1984) Application of multiple echoes energy measurements for evaluation of sea bottom type. *Oceanologia*, **19**, 61–78.

Orlowski, A. (1989) Application of acoustic methods to correlation of fish density distribution and the type of sea bottom. *Proceedings of the Institute of Acoustics*, **11**(3), 179–85.

Parker, N.C., Giorgi, A.E., Heidinger, R.C., Jester, D.B., Jr., Prince, E.D. and Winans,

G.A. (eds) (1990) *Fish-Marking Techniques*, American Fisheries Society, Bethesda, Maryland, USA.

Petitgas, P. (1993) Geostatistics for fish stock assessments: a review and an acoustic application. *ICES Journal of Marine Science*, **50**, 285–98.

Randa, K. (1984) Abundance and distribution of 0-group Arcto-Norwegian cod and haddock 1965–1982, in *Reproduction and recruitment of Arctic cod*, Proc. 1st Russian–Norwegian Symp., Leningrad, 26–30 September 1983 (eds O.R. Godø and S. Tilseth), Institute of Marine Research, Bergen, Norway.

Ricker, W.E. (1948) Methods of estimating vital statistics of fish populations. *Indiana University Publications Science Series*, (15), 1–101.

Ricker, W.E. (1975) Computation and interpretation of biological statistics of fish populations. *Bulletin of the Fisheries Research Board of Canada*, (191), 1–382.

Robinson, B.J. (1984) Calibration of equipment. Subject group C. *Rapports et Procès-Verbaux des Réunions. ICES*, **184**, 62–7.

Roe, H.S.J. and Griffiths, G. (1993) Biological information from an acoustic Doppler current profiler. *Marine Biology*, **115**, 339–46.

Saville, A. (1971) The distribution and abundance of herring larvae in the northern North Sea, changes in recent years. *Rapports et Procès-Verbaux des Réunions. ICES*, **160**, 87–93.

Sawada, K. and Furusawa, M. (1993) Precision calibration of echo sounder by integration of standard sphere echoes. *Journal of the Acoustical Society of Japan (E)*, **14**, 243–9.

Scott, J.S. (1982) Selection of bottom type by groundfishes of the Scotian Shelf. *Canadian Journal of Fisheries and Aquatic Sciences*, **39**, 943–7.

Seber, G.A.F. (1973) *The Estimation of Animal Abundance and Related Parameters*, Griffin, London.

Simmonds, E.J., Williamson, N.J., Gerlotto, F. and Aglen, A. (1992) Acoustic survey design and analysis procedure: a comprehensive review of current practice. *ICES Cooperative Research Report*, (187), 1–127.

Stevenson, D.K., Sherman, K.M. and Graham, J.J. (1989) Abundance and population dynamics of the 1986 year class of herring along the Maine coast. *Rapports et Procès-Verbaux des Réunions. ICES*, **191**, 345–50.

Sundby, S. (1991) Factors affecting the vertical distribution of eggs. *ICES Marine Science Symposia*, **192**, 33–8.

Sundby, S., Bjørke, H., Soldal, A.V. and Olsen, S. (1989) Mortality rates during the early life stages and year-class strength of northeast Arctic cod (*Gadus morhua* L.). *Rapports et Procès-Verbaux des Réunions. ICES*, **191**, 351–8.

Suthers, I.M. and Frank, K.T. (1989) Inter-annual distributions of larval and pelagic juvenile cod (*Gadus morhua*) in southwestern Nova Scotia determined with two different gear types. *Canadian Journal of Fisheries and Aquatic Sciences*, **46**, 591–602.

Talbot, J.W. (1977) The dispersal of plaice eggs and larvae in the Southern Bight of the North Sea. *Journal du Conseil International pour l'Exploration de la Mer*, **37**, 221–48.

Thorne, R.E. (1983) Application of hydroacoustic assessment techniques to three lakes with contrasting fish distributions. *FAO Fisheries Report*, (300), 269–77.

Traynor, J.J. and Nelson, M.O. (1985) Methods of the U.S. hydroacoustic (echo integrator-midwater trawl) survey. *International North Pacific Fisheries Commission Bulletin*, (44), 30–8.

Urick, R.J. (1983) *Principles of Underwater Sound*, 2nd edn, McGraw-Hill, New York.

US GLOBEC (1991) GLOBEC Workshop on Acoustical Technology and the Integration of Acoustical and Optical Sampling Methods. Rep. no. 4, Joint Oceanographic Institutions, Washington, DC.

Walsh, S.J. (1992) Factors influencing distribution of juvenile yellowtail flounder (*Limanda ferruginea*) on the Grand Bank of Newfoundland. *Netherlands Journal of Sea Research*, **29**, 193–203.

Walsh, S.J., Koeller, P.A. and McKone, W.D. (eds) (1993) Proceedings of the international workshop on survey trawl mensuration, Northwest Atlantic Fisheries Centre, St. John's, Newfoundland, March 18–19, 1991. *Canadian Technical Report of Fisheries and Aquatic Sciences*, (1911), 1–114.

Weill, A., Scalabrin, C. and Diner, N. (1993) MOVIES-B: an acoustic detection description software. Application to shoal species' classification. *Aquatic Living Resources*, **6**, 255–67.

Westgård, T. (1989) Two models of the vertical distribution of pelagic fish eggs in the turbulent upper layer of the ocean. *Rapports et Procès-Verbaux des Réunions. ICES*, **191**, 195–200.

Wetherall, J.A. (1982) Analysis of double-tagging experiments. *Fishery Bulletin*, **80**, 687–701.

Young, D.A. (1994) *The X Window System: Programming and Applications with Xt*, 2nd edn, Prentice-Hall, Englewood Cliffs, New Jersey.

Chapter five

Geostatistics and their applications to fisheries survey data

Pierre Petitgas

Geostatistics (Matheron, 1971) is an application of probability theory to estimate statistics relating to spatial variables. Geostatistics is applied in two stages. First, the structural analysis characterizes the different aspects of the spatial distribution. Spatial distributions generally have two aspects, one is structured, the other complex and erratic. During this stage, a model is chosen to interpret the data. The second stage involves using the model to derive the estimates.

The structural analysis is the keystone of the method. During this stage the scientist characterizes the observed variable on the basis of data properties. The spatial structure model describes the average correlation between two points in space. The model choice provides a physical interpretation of reality while the data only provide a fragmentary indication of this reality. The move from data description to reality description involves estimating an average, and hence the use of a probability approach.

The mathematical framework used in geostatistics is the one of random functions. The density surface sampled at a certain number of locations is interpreted as one outcome of a stochastic spatial process. The data give information on that particular sampled realization of the process. The model of spatial structure characterizes the underlying process, not the particular sampled realization.

The second phase of geostatistics, estimation, uses a computerized mathematical algorithm named kriging. Kriging estimates the process values by performing a weighted average of the sampled values. In the kriging procedure, the weights assigned to data values are appropriately determined according to spatial structure and sampling configuration. In the kriging

Computers in Fisheries Research.
Edited by Bernard A. Megrey and Erlend Moksness.
Published in 1996 by Chapman & Hall, London. ISBN 0 412 59550 8

procedure, the estimate is given together with its associated error. Kriging usually serves to reconstruct the process at locations unsampled (mapping) or to estimate the process average values over defined areas. Though the estimates depend on the structural model chosen, similar models give comparable results.

In recent years, fisheries scientists have shown great interest in geostatistical tools when analyzing survey data of fish stocks in the purpose of estimating population abundance (Anon, 1991; Simmonds *et al.*, 1992). The geostatistical approach has enabled survey design and data analysis to be reconsidered. As a consequence of the fact that the structure is modeled and used to derive the estimates of variance and abundance, there is no theoretical constraint in geostatistics on the type of survey design to be performed, the only requirement being a good overall coverage of the area. In particular, random sampling is not required. When sampling is random, sample point locations are independent of each other, thus the mean estimate and its precision can be derived from the sample values directly without making any assumption on the fish spatial distribution. In such a scheme, the spatial coverage may be uneven which may result in a lowering of precision. On the contrary, nonrandom sampling with even spatial coverage will provide precise estimate of mean density but precision cannot be so easily computed. In such a scheme, sample point locations will be correlated and thus precision cannot be estimated without inferring a population structural model. Geostatistics is a relevant tool in this instance.

The possibility of characterizing spatial structure of surveyed fish populations and then using this information for optimizing sampling strategy has perhaps been the most important motive for analyzing fisheries survey data with a geostatistical approach. The application of the kriging methodology for deriving an abundance estimate has also raised interest as kriging provides a weighting of the data values according to their location and to the spatial structure in the population and, therefore, the algorithm can be usefully applied when sampling is not homogeneous and when sample points are clustered. Geostatistics is now more widely used for designing surveys at sea, estimating fish population abundance with its measure of precision and for mapping marine resources. Characterization of population structuring is one of the most attractive features of geostatistics for the biologist.

First, this discussion looks at the structural analysis phase. It then shows how the structural model is used for computing estimation error. Attention is paid to deriving the estimation error associated with a given survey design when estimating population abundance. The details of kriging are then examined. The fundamentals are explained using standard linear geostatistics which are applied to a small fishery survey data set. Finally, the need in fisheries for more advanced structural models is discussed.

5.1 STRUCTURAL ANALYSIS – THE VARIOGRAM

The spatial distribution of fish density can be represented as a complex density surface. The variogram is a useful tool to describe the structural properties of this surface. The variogram measures the level of dissimilarity between points as a function of the distance between them. The variogram unit is formally equivalent to a variance. If the variogram curve increases with distance, then closer points are more similar than more distant points and therefore there is spatial structuring. If the variogram is flat, then close points are as different as distant ones and therefore there is no spatial structuring. The variogram curve can be calculated along several directions to test for directional differences in structure (anisotropy).

There are three types of variograms: the regional variogram, the experimental variogram, and the model variogram. The regional variogram is the variogram function that would be obtained if all values of the surface under study were known. This is the variogram of interest as it characterizes the real spatial distribution. The experimental variogram is the estimate of the regional variogram, the function that can be computed from the data. The model variogram is a probabilistic interpretation characterizing two major types of structural properties; the behavior of the model for short distances indicates the roughness of the density surface (small-scale irregularity), and the behavior over larger distances indicates the range of correlation (Figure 5.1).

Variogram modeling is based on probability interpretation. Consider a density surface sampled at a given time with a given sampling grid. At a later time, new observations are taken on the same grid. The underlying phenomenon is the same, but the values will have changed. The density surface that has been sampled each time is interpreted as one outcome of a random function which characterizes the underlying phenomenon. The difference between the sample and the model is similar to the difference between a single realization of a stochastic simulation and the model used to generate it.

The procedures for calculating variograms are described in Matheron (1971), Journel and Huijbregts (1978), Isaacks and Srivastava (1989), or Armstrong *et al.* (1992). The main points are summarized below.

5.1.1 Experimental variogram

The variogram measures the mean square value of the difference between two points separated by a distance h. The computation can be performed as follows. Take two sample data points. Compute the norm and the direction of the vector they define. Also compute the square of the difference between their values, c^2. Repeat this for all data pairs successively, then group the

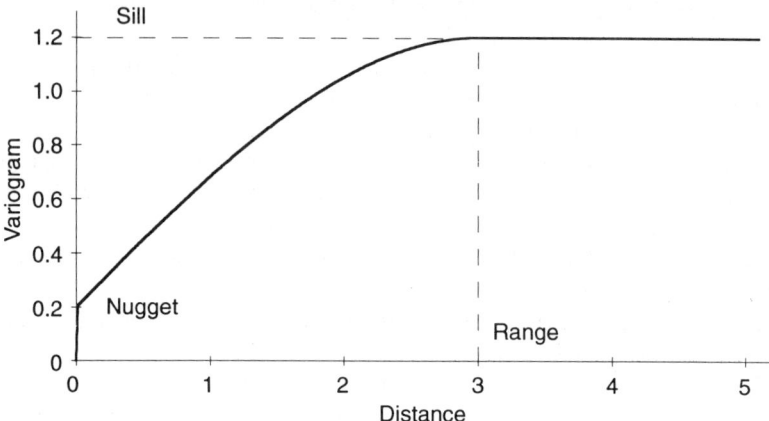

Fig. 5.1 Theoretical shape of a variogram showing the main informative structural parameters; sill, range, and nugget.

directions in angle classes and, for each angle class group, the norms in distance classes. For each distance class of each direction class, compute the average of the c^2 values and divide by 2. Thus the variogram is computed as a function of distance and direction. The class width (generally called the lag distance) is chosen to compute the average c^2 from a sufficient number of pairs. The experimental variogram formula can be written as

$$\gamma^*(u, h) = \frac{1}{2n(u, h)} \sum_i (f(x_i) - f(x_i + h))^2 \tag{5.1}$$

where $\gamma(h)$ denotes the variogram with the asterisk meaning that it is the experimental variogram, $f(x_i)$ denotes the data value measured at point x_i, $n(u, h)$ denotes the number of data pairs for direction u and distance h, and i is the index of the data. Because the variogram value for distance h measures a variance and because of the 2 in the denominator, the variogram unit is often called semivariance. The variogram values are also often named in the same way. The term variogram is often used restrictively to denote the variogram curve and not its values.

The experimental variogram can always be computed and always provides information about spatial structure. Interpretation of its behavior is generally instructive. Contemporary graphical tools will facilitate structural analysis at this stage by permitting simple graphical displays of the data.

Experimental variograms, $\gamma^*(h)$, may be erratic for fisheries data although there is structuring for different ranges of data values. This is particularly true for acoustic pelagic data as the data histogram may be very skewed and high values may occur close to very low ones. In practice, there are different ways of dealing with this problem: Cressie and Hawkins (1980) proposed a different formula for the experimental variogram, and Journel and Huijbregts (1978) proposed normalizing data transformations based on the data cumulative probability distribution. Extreme skewness in the data physically means that there are high density patches very concentrated in space. If this cannot be described by a single model for all the ranges of data values, nonlinear geostatistics (Section 5.4) could be a relevant solution to the problem (Petitgas, 1993b).

5.1.2 Variogram models

The variogram model corresponds to a probabilistic interpretation of the data. The use of mathematically appropriate models is required to ensure that calculated variances are positive (Table 5.1).

The variogram models currently used (Journel and Huijbregts, 1978) describe physical properties of the spatial distributions. These properties are characterized by the behavior of the variogram at short distances (i.e. at the origin for $h = 0$) and at large distances to show whether there is a sill or no sill. Examples of the variogram models are illustrated in Figure 5.2.

The existence of a sill means that there is a maximum level of heterogeneity. Theoretically, this means that the random function model has a variance which is measured by the sill. However, the sill is not always equal to the variance shown by the function over a given zone. The sill is associated with another parameter, the range, which is the distance at which the sill is reached. Two points separated by a distance greater than the range are uncorrelated while closer points are correlated. The range gives a quantitative measure of the average zone of influence around each point. This parameter can be physically related to the average diameter of 'patches'.

Table 5.1 Equations of currently used variogram models. Parameter C is the sill and the r is the range

Model type	Variogram equation: $\gamma(h) =$
Spherical	$C(1.5(\lvert h \rvert / r - 0.5(\lvert h \rvert^3 / r^3))$ if $\lvert h \rvert \leqslant r$ C if $\lvert h \rvert \geqslant r$
Exponential	$C(1 - \exp(-\lvert h \rvert / r))$
Gaussian	$C(1 - \exp(\lvert h \rvert^2 / r^2))$
Power	$C\lvert h \rvert^a$

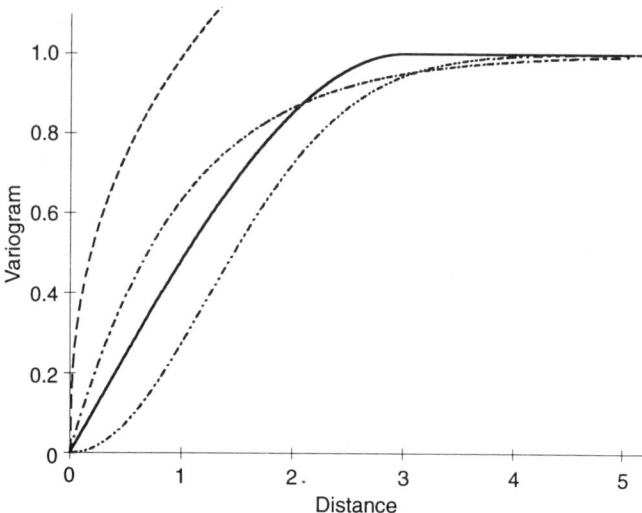

Fig. 5.2 Shapes of the variogram models currently used. The sills equal unity and the practical ranges equal 2.5. Distance and variogram values are in arbitrary units: (——) spherical; (–·–·–) exponential; (–···–) gaussian; (– – –) power (0.2).

Variogram models without sills relate to random functions without variance (infinite variance): the greater the distance, the greater the variance between the points, without limit. However, it is always possible to calculate the variance for a given zone if the variogram is known. It is the average of the variogram values for the distances applied to the surface.

Near the origin, the analytical behavior of the variogram model curve will indicate the degree of irregularity. Vertical tangent behavior indicates very high irregularity while horizontal tangent behavior indicates great regularity. Between the two, linear behavior indicates moderate irregularity (Figure 5.2 and Table 5.2).

The nugget is a discontinuity with amplitude C_0 at the origin of the variogram (Figure 5.1: $\gamma(0) = 0$; $\gamma(h) = C_0 > 0$ as soon as $h > 0$). It is a random component of the spatial distribution with variance C_0. The nugget has three physical interpretations which cannot be distinguished in practice. These are a purely random component of the spatial distribution, and/or a measurement error, and/or a sum of structures which have ranges smaller than the sampling mesh grid. A variogram model can be the sum of different nested models.

In the geostatistical approach of Matheron (1989a), model fitting is considered as a characterization of reality and not as a statistical fitting

Table 5.2 Currently used variogram models and their physical characteristics

Model type	Sill	Behavior at origin	Modeled irregularity
Spherical	yes	linear	medium
Exponential	asymptotic	linear	medium
Gaussian	asymptotic	parabolic (horizontal tangent)	very smooth
Power (h^a)			
$0 < a < 1$	no	increasingly vertical tangent as a tends towards 0	very irregular
$a = 1$	no	linear	medium
$1 < a < 2$	no	increasingly horizontal tangent as a tends towards 2	smooth

problem. The model is considered a reasonable approximation of reality that enables later a consistent estimation. First, the model type is selected for its properties at large and small distances. This choice corresponds to a physical interpretation of the spatial process from which the data have been sampled. The model parameters are then chosen so as to fit the model to the experimental variogram. Matheron (1971) supplies theoretical proof that the variability between regional variograms for different realizations of the same underlying process is small for short distances, thus inference of a population model is possible. Also, Matheron (1989a) provides experimental proof that similar models give similar kriging estimates and variances, thus small differences in model parameters have little influence. Cressie (1991) gives an extensive bibliography for fitting variogram models by least-squares procedures and Pelletier and Parma (1994) fitted a variogram model for halibut densities using weighted least squares. There are also two simple procedures to check whether a model is or is not adapted to describe the data. The dispersion variance over the entire field (Section 5.2) predicted with the variogram model should not be too far from the data variance (Matheron, 1971). Cross-validation of the data values by kriging provides a way to measure how the model and the kriging reproduce the data (Journel and Huijbregts, 1978; Isaaks and Srivastava, 1989).

5.1.3 Anisotropies

Anisotropy means that not all spatial directions are equivalent in their characteristics. So an anisotropic variogram has different characteristics in different directions. Journel and Huijbregts (1978) consider two types of anisotropy, geometric and zonal, with reference to the variogram models with sills.

In geometric anisotropy variogram models, the sill is the same in all directions but the range varies elliptically according to direction, i.e. all directions are equally nonhomogeneous, but the shape of the aggregates is elliptical rather than circular as in the isotropic case. With zonal anisotropy, the variance of the function, or the sill, is a function of direction, i.e. the phenomenon is more heterogeneous in certain directions. The two anisotropies can be combined.

Geometric anisotropy is characterized by an angle of rotation and has an anisotropic coefficient X for the direction of anisotropy, and a Y coefficient for the perpendicular direction. The angle defines the rotation needed to bring the abscissa axis onto the major (or smaller) axis of the ellipse of ranges. The coefficients indicate the expansion or contraction needed on the major and minor ellipse axes to change the ellipse into a circle. For example, suppose that in direction $45°$ the range is twice that in the perpendicular direction, then the rotation is $45°$ and the coefficients are 1 for X and 2 for Y or, equivalently, 0.5 for X and 1 for Y.

Zonal anisotropy is modeled by a nested structure model: there is one model more in the direction showing higher variance. Journel and Huijbregts (1978) recommend using an isotropic model where none of the coefficients along X and Y are zero and adding to it in the zonal anisotropic directions other models. These will have their coefficients X or Y equal to zero. There will be no abrupt discontinuity in structure with varying direction because all directions will share the same isotropic variogram model. The directions showing zonal anisotropy will then be characterized by a specific model that adds in this direction to the isotropic structure. It is preferable to avoid models made up of two nested models where each one is valid only in one sector (Journel and Huijbregts, 1978).

5.2 USE OF THE VARIOGRAM – ESTIMATION OF VARIANCE

The variance between values is a function of their autocorrelation. Thus the variance of the density surface can be computed in any given area by using the variogram and the geostatistics formulae. The variance is always computed over a given area and will depend on the dimensions of the area as well as on the variogram model.

In recent years, fisheries scientists involved in survey design and abundance estimation of fish stocks have shown great interest in the geostatistical formulae for variance estimation. The geostatistical estimation variance can be used for quantifying the precision of stock abundance estimates (Anon, 1991; Petitgas, 1993a). Scientific surveys at sea are primarily undertaken to estimate fish stock abundance. The estimation of abundance amounts to estimating the average fish density over the area surveyed. Precision of the mean estimate is required. Geostatistics provides an alternative to the random sampling theory when the sampling scheme is not performed randomly. In random sampling theory, the random positioning of individual observations ensures that, even though there may be spatial structuring in the population, there will be no correlation in the sample among the data values (Cochran, 1977). The estimate of variance is free from assumptions on the spatial distribution and is computed from the data values directly. It is design-based. For any other type of sampling, the individual observations will not be independently sampled from each other and there will be correlation among observations in the sample set. The estimation of variance in this case requires that one assumes a model for the spatial correlation structure in the population. Thus geostatistics is a sound methodology for deriving a variance estimate in this case. It is a relevant methodology for systematic designs and any type of zig-zag transect designs (Simmonds *et al.*, 1992). The geostatistical variance estimate is model-based.

Matheron (1971) distinguishes two types of variance, the dispersion variance and the estimation variance. The dispersion variance for a given zone V (γ_{VV}) is the variance of all the values of the density surface found in V. It is a theoretical value given by the model for zone V. It is the variance of the random function over zone V. It equals the average of the variogram values for all the distances in V and is denoted γ_{VV}. γ_{VV} can also be estimated by a set of n random samples, z_i in V, by the usual formula:

$$\gamma_{VV} = \frac{1}{n-1} \sum_{i=1}^{n} (z_i - \bar{z})^2$$

A fitted variogram model will be well scaled if its γ_{VV} value is close to the data variance. Thus the variogram sill does not necessarily equal the data variance.

The estimation variance is the variance of the difference between the zone mean and the estimated value, for many repeated surveys performed over the area at the same sample points. The different surveys are assumed to sample the same underlying stochastic process so that the zone mean changes among surveys and thus the difference between the zone mean and the estimated mean changes among surveys. The estimation variance is the variance of these differences. It is a mean-square error that could be called the interpolation variance.

The estimation variance is expressed as a function of the variogram. Suppose we estimate the mean for zone V and that the chosen estimator is the simple average of the data, then the variance of this estimation is (Matheron, 1971):

$$\sigma_E^2 = 2\gamma_{iV} - \gamma_{VV} - \gamma_{ij} \qquad (5.2)$$

where

$$\gamma_{VV} = \frac{1}{V^2} \int_V \int_V \gamma(|x-y|)\, dx\, dy$$

is the mean of the variogram values for all the distances in zone V,

$$\gamma_{ij} = \frac{1}{n^2} \sum_i \sum_j \gamma(|x_i - x_j|)$$

is the mean of the variogram values for all the distances between the samples (i and j are the data indexes), and

$$\gamma_{iV} = \frac{1}{nV} \sum_i \int_V \gamma(|x_i - y|)\, dy$$

is the mean of the variogram values for all the distances between the sample points and all the points of zone V.

Equation (5.2) has three terms. To be computed, these require the use of a variogram model. The estimation variance depends on the variogram model and the shape of zone V (term γ_{VV}), the position of the samples in relation to each other (term γ_{ij}), and the location of the samples in relation to the limits of the zone (term γ_{iV}). The more regular the structure (the smoother the density surface) and the tighter the sampling, the smaller the variance.

In equation (5.2) the sample locations are fixed in the zone under study. The variance is the one over repeated surveys performed on the same fixed stations. When the samples are randomly located, the estimation variance is obtained by averaging equation (5.2) over all possible sample point locations in the zone V. For a random sampling of n observations, the estimation variance becomes

$$\sigma_E^2 = \frac{\gamma_{vv}}{n} \qquad (5.3)$$

This is similar to the variance formula in the random sampling theory. There is no theoretical difference between the two methods. Here, the model is used to estimate the dispersion variance where, as in the random sampling theory, it is estimated directly from the sample values.

Equations (5.2) and (5.3) do not take into consideration the estimation errors made at the limits of the fish stock distribution zone. These errors affect the mean and are another source of variability of the mean estimate. A new term, a geometric error term, should thus be added to the estimation variance calculated with equations (5.2) and (5.3). Matheron (1971) provides the formula of the error term to be added when the sampling scheme is a systematic survey. In practice, the value of this term is often low when the sampling effort is sufficient.

The estimation variance is a linear function of the variogram. It depends greatly on the behavior at the origin of the variogram model (Matheron, 1971). Slightly different models with similar behaviors will give similar estimation variances (Matheron, 1989a).

The estimation variance in equation (5.2) does not depend on the sample values directly but on the variogram model and on the position of the data points. Thus, when the structure is consistent in time, this equation can be used to evaluate the precision given by different survey designs, when estimating fish abundance. The estimation variance can be computed for different sampling configurations with the same variogram model to find the design giving lower estimation variance. The sampling grid can thus be adapted to the anisotropy of the structure and to the range of the spatial correlations.

In the early 1960s, the integrals in equation (5.2) were calculated formally for different types of regular sampling designs and variogram models. Charts were derived for practical use and are available in Matheron (1971) and Journel and Huijbregts (1978). The estimation variance values can be read from these charts. Equations (5.2) and (5.3) can also be computed by computerized numerical integration, whatever the sampling configuration. The field V is required to be finely discretized. The description of the algorithm for computing the integrals of the variogram can be found in Journel and Huijbregts (1978) or in Deutsch and Journel (1992).

Geostatistical estimation variances have been extensively compared with other approaches for computing variance for acoustic fish surveys (Anon., 1991; Simmonds *et al.*, 1992). The geostatistical variance is not necessarily the smallest one. This depends on the range of the correlation structure, on the dimensions of the surveyed area, and on the intensity of sampling. Simmonds and Fryer (1993) have used simulations to investigate how these factors interact. Foote (1993) provides a case study on repeated surveys for herring where survey coverage of the area greatly influences the estimation variance. Pelletier and Parma (1994) provided a case study where a longline survey design for halibut is optimized based on the variogram structure.

In acoustic surveys, there is continuous sampling along the ship's sailing track. Thus when the survey is made up of parallel transects, these can be treated as sample units and the abundance estimation can be performed in one dimension using estimates of fish biomass per transect. Jolly and Hampton (1990) recommended that the transect positions be randomized so that the random sampling theory can be used to estimate variance. Petitgas (1993a) proposed the use of a simple 1-D geostatistical procedure for estimating the variance of the abundance estimate for surveys made up of regularly spaced parallel transects.

5.3 USE OF THE VARIOGRAM – KRIGING

Kriging is a method of interpolation. From values recorded at the sample points it is required to estimate the density surface at the nodes of an interpolation grid. A map will then be derived using a contouring procedure. The sampling grid is not necessarily the same as the interpolation grid. As for any interpolation method, kriging allows us to move from a potentially irregular sampling configuration to a regular configuration which will automatically provide a map. Kriging allows us to estimate, on the interpolation grid, the values of the underlying process. The values estimated in this way correspond to the average values of the random function conditional to the observed values. It must be noted that polynomial adjustment methods do not estimate exactly the same density surface (Anon., 1991; Foote and Stefansson, 1993).

Kriging uses the values of neighboring data points to estimate the values of the density surface at unsampled points. The neighborhood is the area around the point to be estimated in which are found the informative data to be used for the estimation. The kriging estimator is a weighted average of the neighborhood data values. The kriging weights depend on the variogram and the sampling configuration. In linear estimation (linear kriging) the kriging weights are independent of the data values, but in nonlinear estimation (disjunctive kriging) the kriging weights also depend on the sample values.

The kriging weights are determined by a procedure of optimization. At a point x_o we estimate the value of the underlying process by the weighted mean of neighboring sample points

$$z^*(x_o) = \sum_{i \in \eta} \lambda_i z(x_i)$$

where the asterisk denotes the estimator, η denotes the neighborhood and i is the data index. The estimation variance at x_o is expressed as a function of the weights λ_i and of the variogram (or equivalently of the covariance which can be deduced from the variogram). The kriging weights are those that minimize the estimation variance. The minimum is called kriging variance. To find this minimum, the estimation variance is successively differentiated relatively to each of the weights λ_i and equalled to zero. The kriging weights are thus obtained by solving a linear system of equations, called the kriging system, which is written in matrix form as follows (Matheron, 1971):

$$[\gamma_{ij}] \cdot [\lambda_i] = [\gamma_{io}] \tag{5.4}$$

where the γ_{ij} is the semivariance (variogram value) between points x_i and x_j, λ_i is the kriging weight assigned to point x_i and where γ_{io} is the semivariance between point x_i and point x_o, and where the data indexes i and j apply only on the points belonging to the neighborhood. $[\gamma_{ij}]$ is a matrix with n lines and n columns assuming that there are n sample points in the neighborhood, $[\lambda_i]$ and $[\gamma_{io}]$ are matrices with one column and n lines.

In addition to the density surface map, kriging allows us to derive the map of kriging variance at each point. This is a summary of the quality of the estimation obtained with the sampling configuration and the high-lighted spatial structure. Kriging is the only method of interpolation that gives a map of estimation accuracy.

Kriging is a general term denoting the fact that the weights are chosen to minimize variance. Different kriging systems can be written depending on what kind of estimation is being performed, and equation (5.4) is then transformed appropriately. First, it may be required to estimate the values of the process at the points of the interpolation grid as with equation (5.4) – point kriging – or an estimate of the mean value in each block v of the interpolation grid may be required – block kriging. In a block kriging system, the right-hand side vector $[\gamma_{io}]$ of equation (5.4) is replaced by the vector $[\gamma_{iv}]$ where γ_{iv} is the average variogram value between sample point x_i and all points of the block v.

Secondly, when computing the weights λ_i that minimize variance it is possible to consider different constraints that the λ_i must satisfy. This uses an algorithm of optimization under constraints. There are different kinds of kriging systems according to the constraints used. For instance, when the entire zone mean is not known, but estimated, as is generally the case in fisheries, it is judicious to impose the condition

$$\sum_i \lambda_i = 1$$

on the weights. This ensures that the estimator is unbiased and kriging with this constraint is called ordinary kriging. Kriging without this con-straint, as with equation (5.4), is called simple kriging. The ordinary point kriging system is deduced from the simple point kriging system of equation (5.4) by adding one line and one column corresponding to the constraint, as follows (Matheron, 1971):

$$\begin{bmatrix} [\gamma_{ij}] & [1]' \\ [1] & 0 \end{bmatrix} \begin{bmatrix} [\lambda_i] \\ \mu \end{bmatrix} = \begin{bmatrix} [\gamma_{io}] \\ 1 \end{bmatrix} \tag{5.5}$$

where μ is a Lagrange multiplier, $[1]$ denotes a line vector of n unity values and $[1]'$ is a column vector of n unity values. In comparison to the kriging variance in simple kriging, the kriging variance in ordinary kriging is inflated by a term corresponding to the fact that the process mean is unknown (Matheron, 1971). More constraints can be imposed on the esti-mator. For instance, the estimator may be required to respect the shape of the spatial distribution of another variable. More lines and columns then only need to be added into the kriging system. This is called kriging with external drift. Of course, the more constraints, the more inflated will be the kriging variance be.

Thirdly, depending on how the nugget is treated in the kriging system, this component can be filtered. Consider a variogram model made of two components, a nugget $COd(h)$ ($d(0) = 0$; $d(h) = 1$ if $h > 0$; CO is the nugget value) added to a structured component $g(h)$ ($g(0) = 0$). The right-hand side vector of equation (5.4) writes $[gio] + CO[dio]$. If the column vector $[dio]$ is omitted in the right-hand side, then kriging filters the nugget and the structured component is estimated only. The kriged density surface is smoother and the kriging variance is reduced from the nugget value. In doing so, the nugget is interpreted as an unwanted random noise as, for instance, a measurement error additive and independent of the sample values. But if the nugget is interpreted as being a characteristic of the fish spatial distribution it should not be omitted in the right-hand side of the kriging system. More generally, in the case where the sampled process is made of a sum of different nested structures, Matheron (1982) developed a factorial kriging system which enables to estimate each spatial component separately.

Lastly, different neighborhood configurations can be used. Kriging can be conducted in a unique neighborhood, where all the samples are used to estimate each point or block of the interpolation grid. Close and distant data are used and a variogram model is needed for all distances, whether large or small. Kriging can also be carried out for a moving neighborhood, in which case the neighborhood is like a small window that is moved. In this instance, kriging can be compared to a form of weighted moving average. In this case, only the samples close to the point to be estimated are used and only a variogram model for small distances is needed. Ordinary kriging is widely used with a moving neighborhood as the condition imposed on the weights has the physical effect of keeping the estimator close to the (local) neighborhood data average.

As for any interpolation method, neighborhood parameters and mesh size of interpolation grid must be selected. If the grid node distance is too small, the map will appear to be very detailed, but the details will be unrealistic. It is not possible to reproduce the heterogeneity of reality on a smaller scale than the sampling mesh size. It is therefore reasonable to choose a mapping scale that is compatible with the sampling one.

There are generally four neighborhood parameters: the shape and size of the neighborhood, the number of neighborhood data points used for kriging, and a selection criteria according to their location. These can all influence the estimation and their choice is largely dependent on sampling configuration and spatial structure. The important thing is that the estimator be well conditioned by data values. Systematic sampling is appropriate for this requirement.

Foote and Stefansson (1993) provide an extensive comparison of the underlying concepts of kriging and of other interpolation techniques used in fisheries, such as generalized linear models. Simard *et al.* (1992) provide a very detailed application of kriging to shrimp trawl samples.

5.4 MORE ADVANCED STRUCTURAL MODELS THAN THE VARIOGRAM

Not all spatial distributions can be described by variograms and other structural models are needed. The previous sections deal with linear and stationary geostatistics. The two basic assumptions are the homogeneity of spatial correlations, at least for short distances (kriging with a moving neighborhood), and the possibility of studying all the value ranges together. The variogram gives the average measure of variability between two points as a function of the distance between them. This means that the variogram is independent of the specific location of samples and of their values.

Consider a density surface with a maximum in the middle of the zone. If this maximum is randomly located, repeated surveys would show that it could be found elsewhere (otherwise it would always be found at the same location). The model of random occurrence is a stationary model: the same physical process takes place at all points. The fixed occurrence model is nonstationary: all regions do not have the same structural process. Nonstationary geostatistics denote a variety of methods such as, universal kriging (Matheron, 1971), intrinsic random functions of order k (Matheron and Delfiner, 1980) and, to some extent, kriging with external drift (Galli and Meunier, 1987). These methods propose models that characterize a random stationary component of the spatial distribution associated with a nonstationary deterministic component. A simple review of these methods can be found in Armstrong *et al.* (1992).

Universal kriging has been applied to fisheries echo-integration survey data (Sullivan, 1991) and ichthyoplankton net hauls (Petitgas, 1991). In the universal kriging approach, a global trend over the area is estimated then residuals from this trend are estimated. The main problem of this technique is that one cannot estimate both the trend and the residuals from one data set as this leads to a biased variogram of residuals (Matheron, 1971). Sullivan (1991) used depth as an ancillary variable for estimating the trend as it was estimated by regressing the echo-integrated densities on depth. Petitgas (1991) used repeated monitoring surveys to estimate the average trend in time and mapped this trend using the theory of intrinsic random functions of order k.

Assume now that the density surface is made of two components: a relatively smooth surface of low and medium values on which there are patches of high values that have their own structure. The structure of the low values is not the same as that of the high values and different structural models for different parts of the histogram would characterize the regionalization better than a mean model for all ranges of values. The nonlinear models (disjunctive kriging models) specify the structure for each part of the histogram and the structural relations that exist between them. There are two types of disjunctive kriging models (Matheron, 1989b): diffusive models that are related to Markov birth and death processes, and models without border effects that are related to Boolean schemes. In a diffusive model, when going from a high density zone to a low density zone, one traverses on average areas of medium density. In a model without a border effect, one does not necessarily traverse areas of medium density. Rivoirard (1993) provides a full discussion of the disjunctive kriging methodology. Petitgas (1993b) applied to acoustic densities of herring a disjunctive kriging model without border effects.

Other models related to point processes can be of interest when modeling the spatial distribution of schooling fish stocks. Marchal and Petitgas (1993) have analysed acoustic data with this approach.

5.5 PRACTICAL EXAMPLE

The different steps of a typical basic geostatistical study are now illustrated on a small data set. The covariance structure is first modeled, and this information is then used for three purposes. First, the precision of the abundance estimate is derived; then the sampling scheme is discussed. A kriged estimate of mean density is also derived and compared to the simple average of the data. Last, a map is drawn by ordinary point kriging. All calculations were performed using a IBM-PC computer.

5.5.1 Data

The dataset was supplied by the Marine Laboratory of Aberdeen (Scotland) as a test dataset used during the geostatistics course given at Centre de Géostatistique, Fontainebleau (France), for the community of fisheries biologists and statisticians of the International Council for the Exploration of the Sea (Armstrong *et al.*, 1992). The data consist of spawning bed herring egg densities from a fjord west of Scotland. The herring eggs lie in specific sediments which form beds. The sampling is performed by dredging the sediment with an appropriate gear. Before sampling, the area of the spawning beds is known approximately. However, since sediments move from one year to the other and since the shapes of beds are not fixed due to currents and to the spawning behavior of the fish, it is not possible to define the exact location and shape of the bed(s) prior to the survey. Thus the sampling must consider two types of information: geometrical, concerning the limits of the bed(s) and numerical, concerning the egg densities on the bed(s). The constraints linked to the environment and to the work carried out on board are such that sampling cannot be performed according to a well-defined scheme; the scheme performed is neither random nor regular. By its nature, this type of sampling is unorthodox. Many surveys for marine resource assessment show these characteristics in various degrees. Each data value is a density, the number of eggs per meter square of sediment. The coordinates of the sample points are expressed in kilometers. They measure longitudinal and latitudinal distances separating each sample from a given origin in the fjord.

A proportional representation of the data is given in Figure 5.3. The

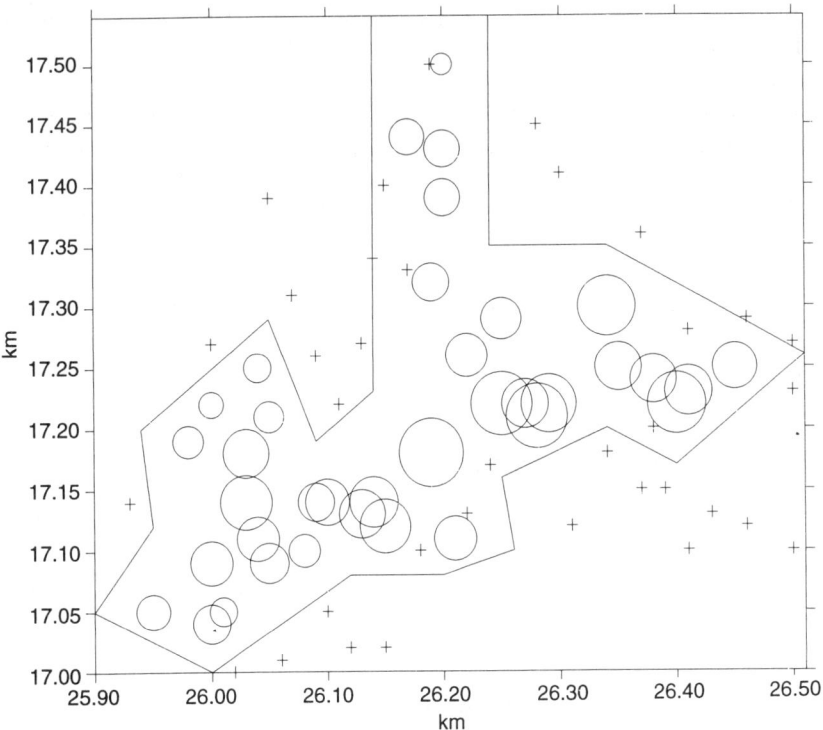

Fig. 5.3 Proportional representation of herring egg densities (eggs m^{-2}) on a spawning bed, in a fjord west of Scotland. Crosses represent zero values. The polygon defines the limits of the spawning bed. Coordinates are expressed in km from an arbitrary origin in the fjord. The data are listed in Table 5.3.

maximum value is represented by a fixed radius circle. All other positive values are represented by circles whose radii are reduced in linear proportion to the maximum. Zero values are represented by crosses. They have a frequency of 0.49 in the dataset. Zero values surround an area composed mainly of positive values. The positive values have a bell-shaped distribution. They are aligned on a main axis whose direction is north-east/south-west. From this main axis, two minor axes develop to the north-north-west. Zero values can be very close to high values at the edges. However, the inner part of the axis does not contain any zero values. On the main axis, there are two aggregates of high values.

The dataset exhibits a geometrical problem. The bed limits are not easily defined for two reasons: precise sampling at the edge is often lacking on the northern side of the bed and, where there are samples (southern side), the limit seems very heterogeneous. Knowledge of the sediment distribution and the biological relation to the presence of the eggs is lacking so that the bed limits cannot be defined precisely.

If the structural analysis is made on the total dataset, the structural information due to the proper shape of the bed (structure of the positive values in the total dataset) will be mixed with the structural information concerning the egg densities on the bed (structure of the values within the bed). If a polygon drawing the limits of the bed can be defined and the inner data selected, the specific structure on the bed can be characterized. This approach is adopted. The abundance estimation will be made in the polygon which defines the bed. Here no information other than the data values can help define the bed limits. Shown in Figure 5.3 is a polygon chosen from the data values. Some zeroes are inside the polygon, thus these are interpreted as interior zeroes, i.e. holes of density on the bed. Other zeroes are interpreted as exterior zeroes (i.e. outside the bed) and these are discarded in the analysis. Sampling outside zeroes enabled a definition of the geometrical limits of the bed. Discarding the exterior zeroes will have no influence on the abundance estimate; however, the estimate of precision will depend on the limits chosen (Section 5.2).

The dataset inside the polygon is reproduced in Table 5.3. The basic statistics of the data inside the polygon are: number of data $n = 46$; simple average $m^a = 963 \, \text{eggs m}^{-2}$; data variance $s^2 = 36.6 \times 10^4$; minimum value = 0; maximum value = 2064.

5.5.2 Structural analysis – variography

No crude anisotropy is identified. An isotropic variogram is computed with a lag distance of 0.05 km and a tolerance of 0.025 km (Figure 5.4). This ensures a reasonable number of data pairs to estimate the variogram values. The experimental variogram shows structuring up to 0.15 km. This

Table 5.3 List of the 46 data inside the zone of interest. X and Y denote, respectively, longitute (km) and latitude (km); Value denotes the observed data value (herring eggs m^{-2}) at coordinate point (X, Y). Coordinates are computed relatively to an arbitary origin

X	Y	Value	X	Y	Value	X	Y	Value
26.10	17.14	1286	26.34	17.30	1786	26.20	17.43	937
26.03	17.18	1352	26.35	17.25	1355	26.40	17.22	1805
26.03	17.14	1588	26.38	17.24	1342	26.14	17.14	1427
26.05	17.09	1039	26.41	17.23	1399	26.25	17.22	1879
26.19	17.18	2064	26.45	17.25	1273	26.29	17.22	1692
26.21	17.11	1216	26.04	17.25	617	26.41	17.28	0
26.13	17.13	1353	25.98	17.19	728	26.38	17.20	0
26.15	17.12	1529	26.09	17.14	968	26.24	17.17	0
26.05	17.21	723	26.04	17.11	1203	26.22	17.13	0
26.08	17.10	798	26.00	17.09	1194	26.19	15.50	0
26.19	17.32	986	25.95	17.05	863	26.18	17.10	0
26.22	17.26	1165	26.20	17.50	372	26.17	17.33	0
26.01	17.05	631	26.17	17.44	884	26.15	17.40	0
26.20	17.39	925	26.28	17.21	1925	26.14	17.34	0
26.25	17.29	1103	26.00	17.22	534			
26.27	17.22	1357	26.00	17.04	1012			

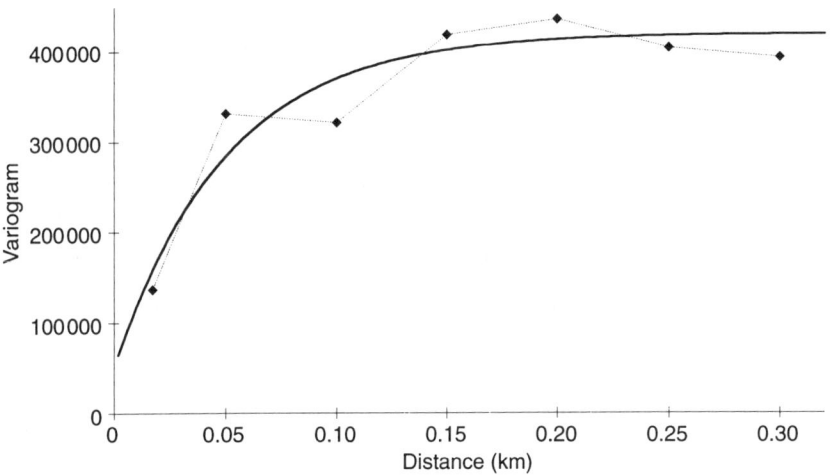

Fig. 5.4 The experimental variogram and its model. The model fitted is: nugget (C_o = 5 × 10^4) + exponential (sill C = 37 × 10^4, practical range r = 0.15 km).

characterizes the average diameter of the two patches described above. The experimental variogram is computed for a maximum distance of 0.3 km which represents half of the length of the bed. For greater distances the increase or decrease of the variogram is influenced by the location of the patches inside the polygon. This is not modeled.

An exponential model with a nugget is fitted to the experimental variogram (Figure 5.4). It is reasonable to want to fit a nugget as we have seen that there are low values very near two high ones on the bed. An exponential model has a linear behavior at the origin which characterizes a medium roughness of the density surface. Models that behave linearly at the origin are a conservative choice when no evidence exists for interpreting the density surface as being very rough or very smooth at small scales. The exponential model is preferred here to the spherical one as it increases faster and is thus more compatible with the experimental variogram. The goodness of fit is decided by eye. The model fitted is a nugget effect added to an exponential. The parameters of the fitted model are: nugget value $C_o = 5 \times 10^4$; practical range of the exponential $r = 0.15$ km; sill of the exponential $C = 37 \times 10^4$.

The sill $(C_o + C)$ is close in value to the data variance. The dispersion variance in the polygon computed with the model is: $\gamma_{VV} = 39.2 \times 10^4$. This is a little higher than the data variance but stays within the same order of magnitude. The model is well scaled to the data variance in the polygon.

5.5.3 Error on the abundance estimate – discussion on the sampling design

The mean density inside the polygon is estimated by using the simple average of the data, $m^a = 963$ eggs m^{-2}. The precision of this estimation is given by equation (5.2): $\sigma_E^2 = 2\gamma_{iV} - \gamma_{VV} - \gamma_{ij}$. The terms γ_{VV} and γ_{iV} are integrals of the variogram for all distances in the polygon V (Section 5.2). The computerized algorithm for these integrations (Journel and Huijbregts, 1978) requires the polygon to be finely discretized. The discretization of the polygon used here is a grid of 20×20 cells, which was found to optimize speed and precision of the computation. The calculation is performed using the model previously fitted. The estimation variance is $\sigma_E^2 = 6303.8$, thus the relative estimation error is $\sigma_E/m^a = 8.2\%$.

What would the estimation variance be for other sampling schemes? This question can be answered considering that other sampling schemes would give data sets out of which a similar variogram model would be inferred and a similar mean estimated. A sampling scheme made of the same number of 46 stations but randomly located inside the polygon would give a relative estimation error of $\sqrt{(\gamma_{VV}/n)}/m^a = 9.6\%$. A random design for the same sampling effort gives a slightly higher error for the mean estimate.

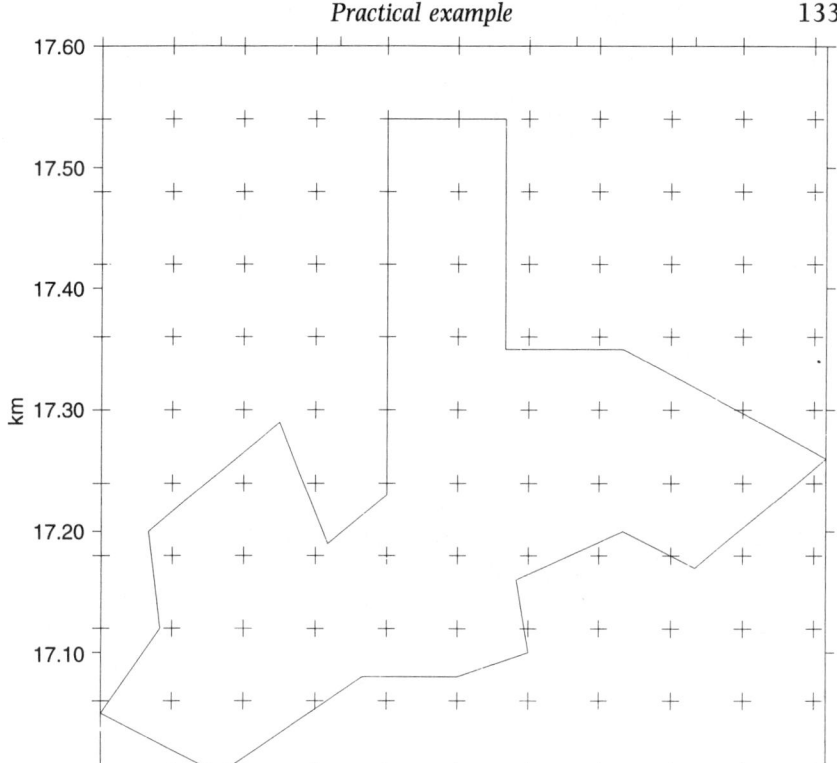

Fig. 5.5 A regular sampling survey for the bed. The samples outside the polygon were not used in the variance computation.

Error for both designs are close in value as the spatial coverage for the scheme performed is uneven (areas in the northern side of the bed stay bare of sample points). A regular grid of sample points is thus expected to improve precision. A grid with a mesh size $0.06\,\mathrm{km}^2$ has been generated (Figure 5.5) whose origin is the coordinate point ($x = 25.9$, $y = 17.0$). This scheme has 36 stations regularly spaced inside the polygon. The relative estimation error calculated for this scheme using equation (5.2) is $\sigma_E/m^a = 8.2\%$. This regular scheme gives the same accuracy as previously obtained, but with ten fewer stations.

The structural analysis, the calculation of the dispersion variance and of the estimation variance were performed by using the public domain software EVA (Petitgas and Prampart, 1993) for an IBM-PC. This software works under Windows and makes extensive use of graphics for visualizing and analyzing the data spatially.

Table 5.4 Summary statistics of the abundance estimation. m^a denotes the data simple average and σ_E^2 its associated estimation variance; m^k denotes the block kriged mean and σ_k^2 its associated estimation variance.

Survey design	Number of data points	Estimate of zone mean (eggs m^{-2})	Relative error for the zone mean (%)
Uneven (performed)	46	$m^a = 963$ $m^k = 944$	$\sigma_E/m^a = 8.2$ $\sigma_k/m^k = 7.0$
Random (tested)	46	–	$\sigma_E/m^a = 9.6$
Regular grid (tested)	36	–	$\sigma_E/m^a = 8.2$

5.5.4 Zone mean derived by kriging

In the previous section, the estimation variance of the data simple average was computed. No kriging was involved. A kriging algorithm will now be used. The zone mean over polygon V is estimated by ordinary block kriging over the entire area V. It is necessary to solve the kriging system of equation (5.5) where the right-hand side (see Section 5.3) is the column vector

$$\begin{bmatrix} [\gamma_{iV}] \\ 1 \end{bmatrix}$$

The average variogram value γ_{iV} between sample point x_i and all points of polygon V is computed using the same polygon discretization grid of

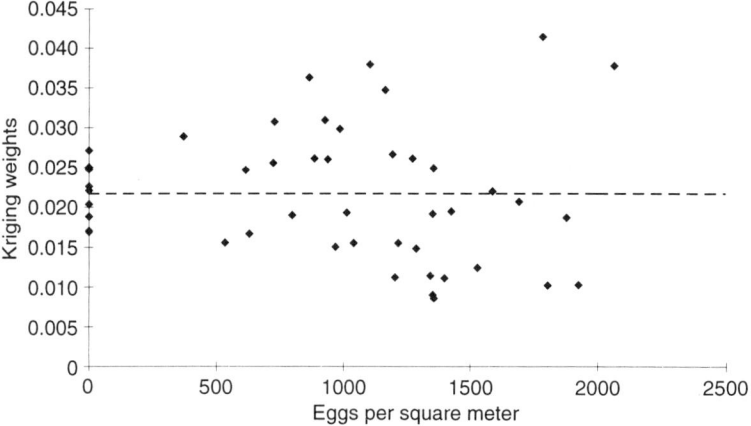

Fig. 5.6 Plot of kriging weights against data values. The weights are obtained by applying ordinary block kriging over the entire polygon. The dashed line represents the weight assigned to each sample value in the data simple average.

20 × 20 cells defined previously. The kriging system has $n + 1 = 47$ lines and the same number of columns.

The zone kriged mean is $m^k = 944\,\mathrm{eggs\,m^{-2}}$. The kriging variance is $\sigma_k^2 = 4351.6$. The kriging relative error is thus: $\sigma_k/m^k = 7.0\%$. The kriged mean is slightly smaller than the data simple average and is more precise. The kriged estimate is optimal, conditional to the fact that the structural model is a good approximation of reality. A summary of the different estimation results is presented in Table 5.4.

The kriging weight values can be compared to the value of $1/n$ $(= 2.15 \times 10^{-2})$ which is the weight assigned to each point when computing the simple average (Figure 5.6). The weights assigned to many medium values are smaller than $1/n$ and this may explain why the kriged mean is lower than the data simple average. The kriging weights are spa-

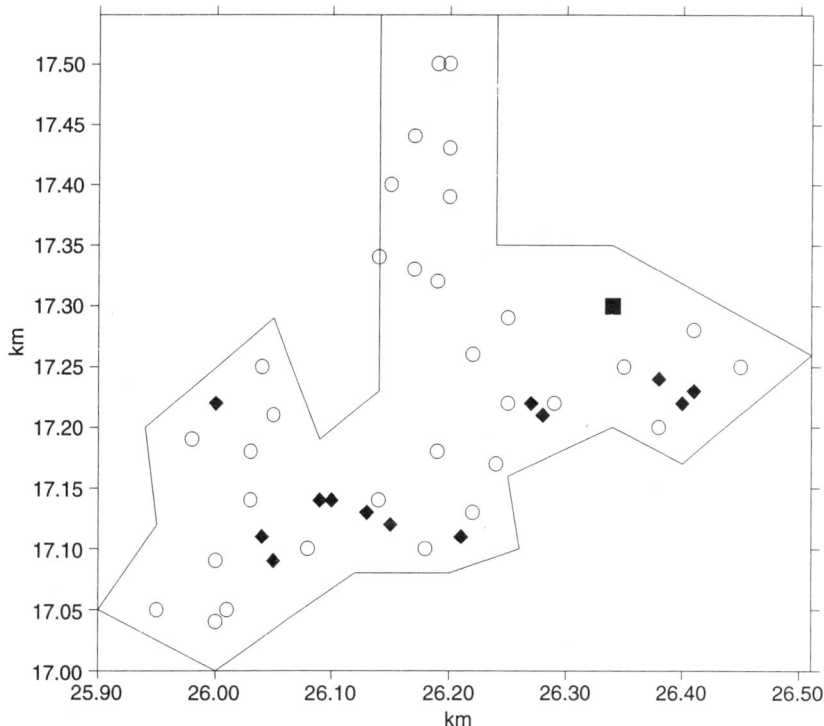

Fig. 5.7 Spatial representation of the kriging weights obtained by applying ordinary block kriging over the entire polygon. ○ represents weights in the range 0.0163–0.0271 (kriging weights close to $1/n$); ◆ represents weights in the range 0.0–0.0163 (kriging weights lower than $1/n$); ■ represents weights in the range 0.0271–1.0 (kriging weights higher than $1/n$), where $n = 46$ is the number of data points.

tially represented to illustrate how the kriging algorithm assigns weights according to spatial structure and relative positions of data points (Figure 5.7). Most of the sample points that are evenly distributed are assigned weights that are close in value to $1/n$. A few points in the center of clustered points are assigned weights lower than $1/n$. One point standing alone is assigned a weight higher than $1/n$. Kriging performs an optimal weighting of the data points conditional on the variogram model and the data locations.

When n is large and the surveyed area is evenly covered, the kriging weights tend to be close to $1/n$. In this instance, the difference between the zone kriged mean and the data simple average is expected to be small. This has been reported for an intense acoustic zig-zag survey on a herring shoal (Anon., 1991, Appendix B). The sample points which were assigned very different weights were the few points standing at the edges of the zig-zag transects where sample points were inevitably in small clusters.

5.5.5 Abundance estimation with a quasistationary variogram

In the present case study, the variogram model is a valid characterization of the spatial structure in all ranges of distances, in other words correlations are assumed stationary throughout the area in all ranges of distances. The zone mean was estimated in one step by block kriging over the entire polygon V and the kriging variance was the associated estimate of precision. When estimating the zone mean with the simple data average, the associated estimation variance was also estimated in one step with equation (5.2) which involves averaging the variogram values for all distances in the field.

When the variogram model is restricted to small distances enabling only kriging with a moving neighborhood (quasistationarity of Matheron, 1971), and when the survey design is not random nor stratified, then the estimation variance over the entire zone can only be estimated by making approximations. This is true whether kriging is involved or not. Approximations have been proposed by Matheron (1971) in the case of systematic sampling designs and by Journel and Huijbregts (1978) when kriging data sampled with an unorthodox survey design. In the case of fish acoustic surveys over large areas that were not designed following random sampling theory, Simmonds *et al.* (1992) suggest the potential benefits of kriging the fish density surface in the area. These authors also propose discretizing the surveyed area in blocks and analyzing these as if they were uncorrelated strata. All suggestions amount to assuming uncorrelation of local estimation errors, so as to estimate the estimation variance of the entire zone by a simple combination of local estimation variances. Simard *et al.* (1992) estimated shrimp abundance from a trawl survey showing clustered trawl stations by mapping the spatial distribution using kriging with a moving neighborhood and taking the average of the kriged estimates. The authors estimated precision assuming uncorrelation of local estimation errors. Although kriging is interesting for weighting clustered data when estimating abundance, the approximation procedure when estimating precision is not always satisfactory. The validity of the approximation made when combining local error terms as if they were uncorrelated depends on the relation between correlation range and grid node distance (Matheron, 1971). Nonstationary geostatistical models can be used for abundance estimation when stationary models are not adapted to the data characteristics (see Section 5.4).

5.5.6 Mapping by kriging

The kriging procedure used here for deriving a map is point kriging with a moving neighborhood. This is a standard procedure. The mesh size of the interpolation grid and the neighborhood parameters must be chosen. The grid mesh size chosen is 0.025 km. This is about the average distance between nearest samples. Thus the interpolation grid is scaled on the scale of the sampling and no illusory detail at a smaller scale will be mapped. The neighborhood parameters have been chosen after a cross-validation procedure.

In a cross-validation procedure each sample is successively removed from the analysis and estimated by kriging as if unknown, using the samples lying in the neighborhood. Statistics of the errors tell whether the kriging parameters produce a kriging algorithm well scaled to the data. Let $z(x)$ be the sample value at point x and $z^k(x)$ its kriged estimate. Let $\sigma_k(x)$ be the kriging standard error of the estimation of $z(x)$. The average bias should be close to zero:

$$E[z(x) - z^k(x)] = 0$$

The average error ratio should be close to unity:

$$E\left(\frac{|z(x) - z^k(x)|}{\sigma_k(x)}\right) = 1$$

Different neighborhood parameters were tested and the one retained has the following characteristics. The shape is a circle with radius of 0.08 km. This distance is smaller than the correlation range. The number of sample points used in the neighborhood were 4 at the most and 2 at the least. If there were more than 4 samples in the neighborhood, the 4 nearest were used. If there were less than 2 samples no kriging was performed. No directional option in searching for the samples was used. The average bias is 44 and the average error ratio is 0.71. The kriging standard error gives a slight overestimate of the reestimation error.

Figure 5.8 shows the scatter plot of the kriged estimates against the

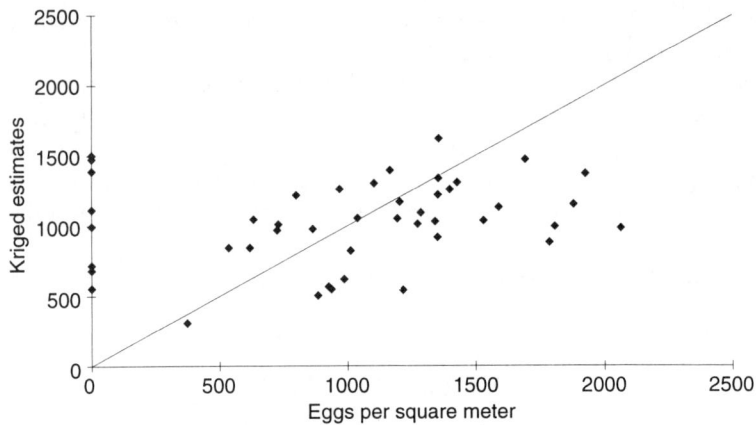

Fig. 5.8 Scatter plot of the cross-validation procedure for the kriging neighborhood parameters chosen.

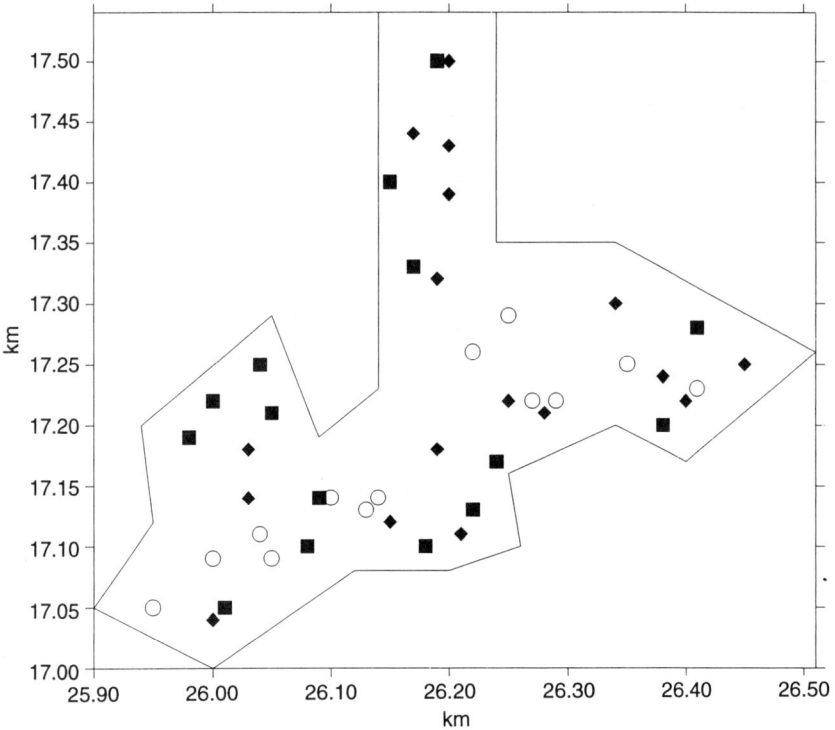

Fig. 5.9 Spatial representation of the cross-validation relative errors for the kriging neighborhood parameters chosen. The relative errors are $[z(x) - z^k(x)]/z(x)$. ◆ represents the range of values -1.5 to -0.05 (overestimation); ■ represents the range of values 0.05 to 1.5 (underestimation); ○ represents the range -0.05 to 0.05 (good estimation).

sample values and Figure 5.9 shows a spatial representation of the relative estimation errors,

$$\frac{z(x) - z^k(x)}{z(x)}$$

High values tend to be underestimated and all the zero values are over-estimated. This results in a small positive overall bias of the cross-validation. As the zeroes are very close to the high values, this effect is more pronounced when a larger neighborhood is chosen and when a greater number of samples are used in the neighborhood. This is why the number of informative data points was restricted. However, this is paid for by a slightly high kriging error as was noticed previously. Except for the highest ones, the values on the main axis of the bed are estimated satis-factorily.

The choice of the neighborhood is not always a simple task. Although helpful, cross-validating the data does not entirely address the problem posed by the neighborhood choice. In ordinary kriging with a moving neighborhood, the kriged estimates are conditioned by the local average of the neighborhood sample points as their kriging weights must sum to unity. Thus, it is tempting to lower the number of sample points used in the

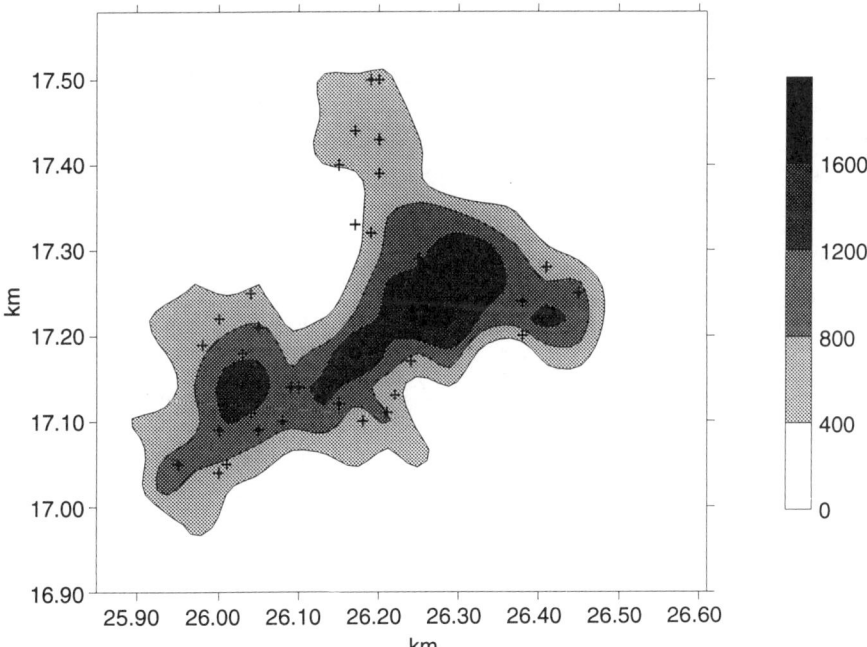

Fig. 5.10 Map of egg densities (eggs m^{-2}) estimated by ordinary point kriging with a moving neighborhood.

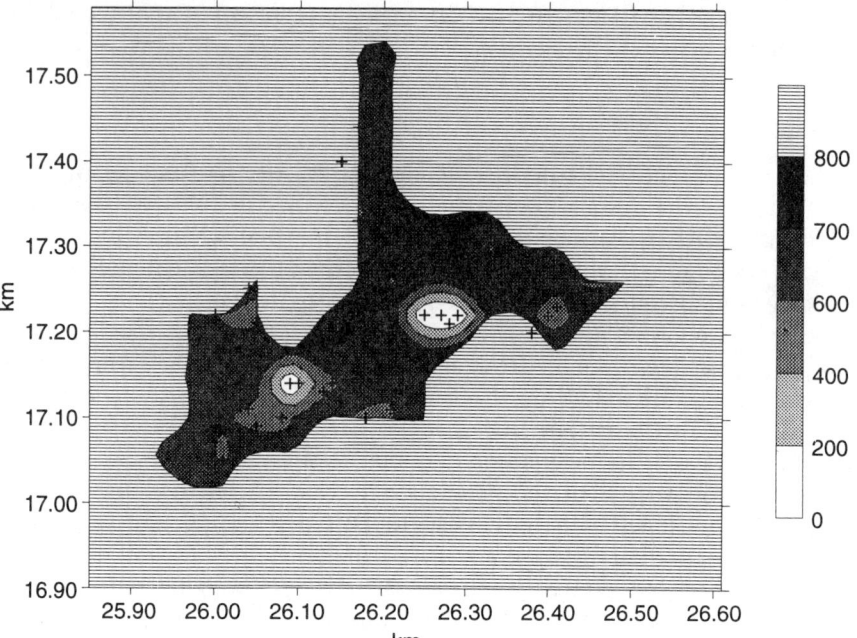

Fig. 5.11 Map of kriging standard errors.

neighborhoods, so as to reproduce the local heterogeneities. There is of course a limit to this temptation which is related to stationarity assumptions over the surveyed area. A small neighborhood increases the estimation variance, particularly when the process mean is poorly estimated (Armstrong *et al.*, 1992).

Kriging is now performed on the grid chosen with the neighborhood parameters described above. The map obtained is shown on Figure 5.10 and the map of the kriging standard errors is shown on Figure 5.11. When the density of sample points is high, then kriging is more precise, and when the density of sample points is low, then kriging is less precise. but the kriging standard error is about equal to the data standard deviation over the largest part of the area.

Cross-validation of the data and kriging of the map were performed by using the public domain software GEO-EAS (Englund and Sparks, 1988) for IBM-PC.

5.6 DISCUSSION AND CONCLUSIONS

Structural analysis is the keystone of geostatistics and is of considerable importance to the fisheries biologist. Graphical displays and point-and-click user-friendly software under the IBM-PC Windows environment should make it more attractive for biologists to analyze the spatial structure of their data and thus gain information on the fish spatial structuring processes.

The basic kriging techniques for mapping marine resources are on their way to being more widely applied. Common mapping commercial software often includes kriging options. The limitation is that the structural analysis is often performed in too automatic a manner that does not allow the user to visualize his data. This may lead to inadequate modeling. Deutsch and Journel (1992) have made available a library of Fortran source code geostatistical algorithms. Public domain geostatistical software is also available and different marine laboratories have developed their own.

The possibilities opened up by geostatistics, to estimate precision of survey abundance estimates of fish stocks when the survey is not randomly designed, has been an important development for fisheries surveys.

Geostatistics is a useful tool for improving survey design, estimating population abundance, and mapping the density field. The geostatistical approach also enables key biological questions to be addressed for fisheries research related to the spatial strategies of habitat occupation by fish populations such as, density dependent structuring processes, or environmentally and behaviorly induced structuring changes. There is a need to understand better the spatio-temporal variabilities of fish spatial distributions and to develop appropriate models. Large scale spatio-temporal datasets of yearly monitoring surveys are now widely available and much information on small-scale school structuring is provided by acoustic surveys. These studies will direct future fishery geostatistics applications more towards the fields of multivariate and nonstationary analysis, temporal variability, nonlinear models and simulations.

REFERENCES

Anon. (1991) Report of the workshop on the applicability of spatial statistical techniques to acoustic survey data, *ICES/CIEM CM91/D:40*.

Armstrong, M., Renard, D., Rivoirard, J. and Petitgas, P. (1992) *Geostatistics for fish survey data*, course publicized by ICES, Centre de Geostatistique, Fontainebleau.

Cochran, W. (1977) *Sampling techniques*, John Wiley and Sons, New York.

Cressie, N. (1991) *Statistics for spatial data*, John Wiley & Sons, New York.

Cressie, N. and Hawkins, D. (1980) Robust estimation of the variogram. *International Journal of the International Association of Mathematical Geology*, **12**, 115–25.

Deustsch, C. and Journel, A. (1992) *GSLIB, Geostatistical software library and user's guide*, Oxford University Press, New York.

Englund, E. and Sparks, A. (1988) *GEO-EAS user's guide*, Environmental Monitoring Systems Laboratory, Office of Research and Development, US Environmental Protection Agency, Las Vegas, Nevada.

Foote, K. (1993) Abundance estimation of herring hibernating in a fjord, *ICES/CIEM CM93/D:45*.

Foote, K. and Stefansson, G. (1993) Definition of the problem of estimating fish abundance over an area from acoustic line-transect measurements of density. *ICES Journal of Marine Science*, **50**, 369–81.

Galli, A. and Meunier, G. (1987) Study of a gas reservoir using the external drift method, in *Geostatistical case studies*, (eds G. Matheron and M. Armstrong), Reidel, Dordrecht, Netherlands, pp. 105–19.

Isaaks, E. and Srivastava, R. (1989) *An Introduction to applied geostatistics*, Oxford University Press, New York.

Jolly, G. and Hampton, I. (1990) A stratified random transect design for acoustic surveys of fish stocks. *Canadian Journal of Fisheries and Aquatic Sciences*, **47**(7), 1282–91.

Journel, A. and Huijbregts, Ch. (1978) *Mining Geostatistics*, Academic Press, London.

Marchal, E. and Petitgas, P. (1993) Precision of acoustic abundance estimates: separating the number of schools from the biomass in the schools. *Aquatic Living Resources*, **6**(3), 211–19.

Matheron, G. (1971) *The theory of regionalized variables and their applications*, Les Cahiers du Centre de Morphologie Mathématique, Fascicule 5, Centre de Géostatistique, Fontainebleau.

Matheron, G. (1982) *Pour une analyse krigeante des données régionalisées*, Note N-732, Centre de Géostatistique, Fontainebleau.

Matheron, G. (1989a) *Estimating and choosing – an essay on probability in practice*, Springer-Verlag, Berlin.

Matheron, G. (1989b) Two classes of isofactorial models, in *Geostatistics*, (ed. M. Armstrong), Proceedings of the third international geostatistics congress, Kluwer Academic Publishers, 309–22.

Matheron, G. and Delfiner, P. (1980) *Intrinsic random functions of order k*, course C-80, Centre de Géostatistique, Fontainebleau.

Pelletier, D. and Parma, A. (1994) Spatial distribution of Pacific halibut (*Hippoglossus stenolepis*): an application of geostatistics to longline survey data. *Canadian Journal of Fisheries and Aquatic Sciences*, **51**, 1506–18.

Petitgas, P. (1991) *Contributions géostatistiques à la biologie des pêches maritimes*, Ph.D. Thesis, Centre de Géostatistique, Fontainebleau.

Petitgas, P. (1993a) Geostatistics for fish stock assessments: a review and an acoustic application. *ICES Journal of Marine Science*, **50**, 285–98.

Petitgas, P. (1993b) Use of disjunctive kriging to model areas of high fish density in pelagic acoustic fisheries surveys. *Aquatic Living Resources*, **6**(3), 201–9.

Petitgas, P. and Prampart, A. (1993) EVA: a geostatistical software for structure characterisation and variance computation, *ICES/CIEM CM93/D:65*.

Rivoirard, J. (1993) *Introduction to disjunctive kriging and non-linear geostatistics*, Oxford University Press.

Simard, Y., Legendre, P., Lavoie, G. and Marcotte, D. (1992) Mapping, estimating biomass and optimizing sampling programs for spatially autocorrelated data: case study of the northern shrimp (*Pandaulus borealis*). *Canadian Journal of Fisheries and Aquatic Sciences*, **49**, 32–45.

Simmonds, E. and Fryer, R. (1993) Survey strategies for structured populations: precision of variance estimators, *ICES/CIEM CM93/D:31*.

Simmonds, E., Williamson, N., Gerlotto, F. and Aglen, A. (1992) Acoustic survey design and analysis procedure: a comprehensive review of current practice, *ICES Coop. Res. Report*, No. 187.

Sullivan, P. (1991) Stock abundance estimation using depth-dependant trends and spatially correlated variation. *Canadian Journal of Fisheries and Aquatic Sciences*, **48**, 1–13.

Chapter six

Pattern recognition

Ferren MacIntyre
and Thomas T. Noji

6.1 INTRODUCTION

Pattern recognition by computers has proven to be much more difficult than optimists once hoped. Tasks which are far more mechanical, and possessed of rigid, well known rules, such as architecture and engineering, are now more realistically described as computer-aided rather than computer-performed operations. The reason is simple: silicon- and carbon-based intelligence behave very differently, and are good at very different things. Neither is a replacement for the other, but learning how to work together, and how best to divide the workload, is proving a difficult challenge.

The standard response to requests for automated pattern recognition for the last 2 decades has been: 'Tell me how you identify your friends, and I'll teach the computer to do it too.' As we write, the fastest Princeton-architecture computers remain mulishly inept at such simple tasks as recognizing handwritten characters, with 100 characters per second pushing the start of the art. (Also called von Neumann architecture, its distinguishing feature is that code and data are treated identically, making it possible to write self-modifying code (an essential in the days of small memory and primitive operating systems). Harvard architecture keeps instructions and data separate.)

We seem to have been pursuing the wrong approach for we cannot, in fact, describe how we recognize our friends, any more than we can describe how we move a finger. It just happens – because the brain is designed for such operations. Face recognition may be an unfair example, since the human brain is so highly optimized for facial recognition that it is able to see faces where no faces are (Gould, 1994), such as the one on the Makapansgat Pebble, lovingly carried 3 miles home by an australopithecine 3 million years ago (Morris, 1994).

One cell in a monkey's brain fires when it sees upright monkey hands

Computers in Fisheries Research.
Edited by Bernard A. Megrey and Erlend Moksness.
Published in 1996 by Chapman & Hall, London. ISBN 0 412 59550 8

(Ornstein and Thompson, 1984). Another recognizes front views of faces (monkey and human), and another, facial profiles (Desimone, 1991). It is an oversimplification to say that these cells are responsible for recognition, for other cells that are not being monitored may also be firing. Also, these cells seem to work by weighting inputs from perhaps 10 000 other neurons, each similarly connected.

Neural networks, which attempt to emulate in silicon the connectivity and structure of biological neurons, may be a better approach than pixel algorithms. Intel hopes that its NI1000 'recognition accelerator', containing 3.7 million transistors and 1 K (1024) 'silicon neurons', will increase character recognition to $40\,000\,\text{s}^{-1}$ (Anon., 1993). Training such a device does not consist of writing stepwise algorithms, but of showing it letters, telling it whether its identification was right or wrong, and letting it adjust the weights it gives to its inputs. This rather magically avoids the problems inherent in recognizing friends. On the down side, it is impossible to tell what it is doing, what the final weights mean, or what it is that the net is really learning. The classical example is a military task involving photos of woods with and without tanks. The net learned to separate photos into two categories easily enough, with all the tanks in one pile, but later investigation showed that the two sets had been taken during different weather conditions and what the net had learned was to sort out shadows – it knew nothing of tanks. Showing such a net something outside its training set is always interesting and likely to be unpredictable. It is only conjecture to suppose at this stage that enough NI1000s, properly interconnected, might find the ability to recognize a hand within their abilities. Then the question becomes, could this same connection also be taught to recognize *Calanus finmarchicus* and separate it from its relatives? There are at present only the vaguest notions of the properties of such neural nets, or the relations between complexity, skill, capacity, sensitivity, resolution, and general utility. They promise much excitement, but the state of the art is still depressing. One of the most advanced topics, because much desired by telephone companies, is automatic recognition of spoken numbers, for which Rahin (1992) reports 50% reliability. With sufficient CPU power – 500 million floating-point multiply-and-add operations – neural nets can now find the eyes in a frontal-face image with about 90% accuracy. Those who achieved this say that it is 'unrealistic to strive to produce a system capable of 100% recognition rates' (Vincent *et al.*, 1992).

If the present achievements of pattern recognition by silicon seem minuscule, keep in mind the disadvantages under which it is working. Figure 6.1 shows one view of our current understanding of how a functional pattern-recognition system is organized. The diagram contains many abstractions, and the purpose of most of the boxes is unknown. Edge detection, which is a fairly serious suite of algorithms for a desktop silicon computer, is handled by the boxes labelled Retina and V1. Not shown in the figure are several important, and indubitably complex, inputs from other systems. Visual circuitry which distinguishes between a moving field which represents motion of the environment, and the same field resulting from motion of the head, and shuts off portions of the system during eye movements, does so by combining input from the proprioceptive system, at a high hierarchical level (Burr *et al.*, 1994). Characteristic sounds predispose us to see certain patterns in certain directions: this requires substantial input from far up the auditory system. We have all had the experience of looking at something (often a reduced-dimension image) and not having any idea of what we were looking at for several seconds. This time lag represents the optical system calling upon other systems for help, most likely the file system, and appealing to an ever widening sphere of less often consulted processors for ideas. People in cultures which do not make graphical representations of humans are usually unable to recognize people in photographs. None of these high-level linkages is indicated in Figure 6.1.

A little-appreciated feature of our vision is the remarkable concentration of receptors in the fovea, and their even more lopsided representation in the brain. Each foveal cell has some four times the cortical allocation that a peripheral cell has (Azzopardi and Cowey, 1993) with the result that 0.01% of our retina gets to use 8% of the visual cortex. The object we are looking at in detail receives 800 times the attention that the same area of background gets. Machine vision does not have this advantage, and must attach equal importance to each pixel.

Figure 6.2 shows a silicon-based pattern recognizer. It will be appreciated that there is room for further development. It is not meant to imply that imitation of the behavior of carbon-based intelligence is the natural route for the development of silicon intelligence, only that silicon has a long way to go to reach the level of complexity that pattern recognition seems to require. It is interesting to note that insects, with very much smaller brains, process visual information very much as do vertebrates. Back-to-back papers in *Nature* suggest that feature extraction in bees (Srinivasnan *et al.*, 1993) and dragonflies (O'Carroll, 1993) might be comparable to, if less extensive than, mechanisms known to exist in mammals. This probably represents parallel evolution – that is, an experimental consensus on the best way to approach the problem – so that it might make very good sense to attempt to duplicate the insect optical processor in silicon.

Workers at the University of Pennsylvania are analyzing 128×128 pixel images on a workstation with a software neural net called NEXUS, containing ca. 1 million units and 100 million interconnections, in which each 'unit' can itself be a complex circuit (Finkel and Sajda, 1994). With this much power in a neural net, they are having some success with Gestalt perception, teaching the net to identify contours and assign them to

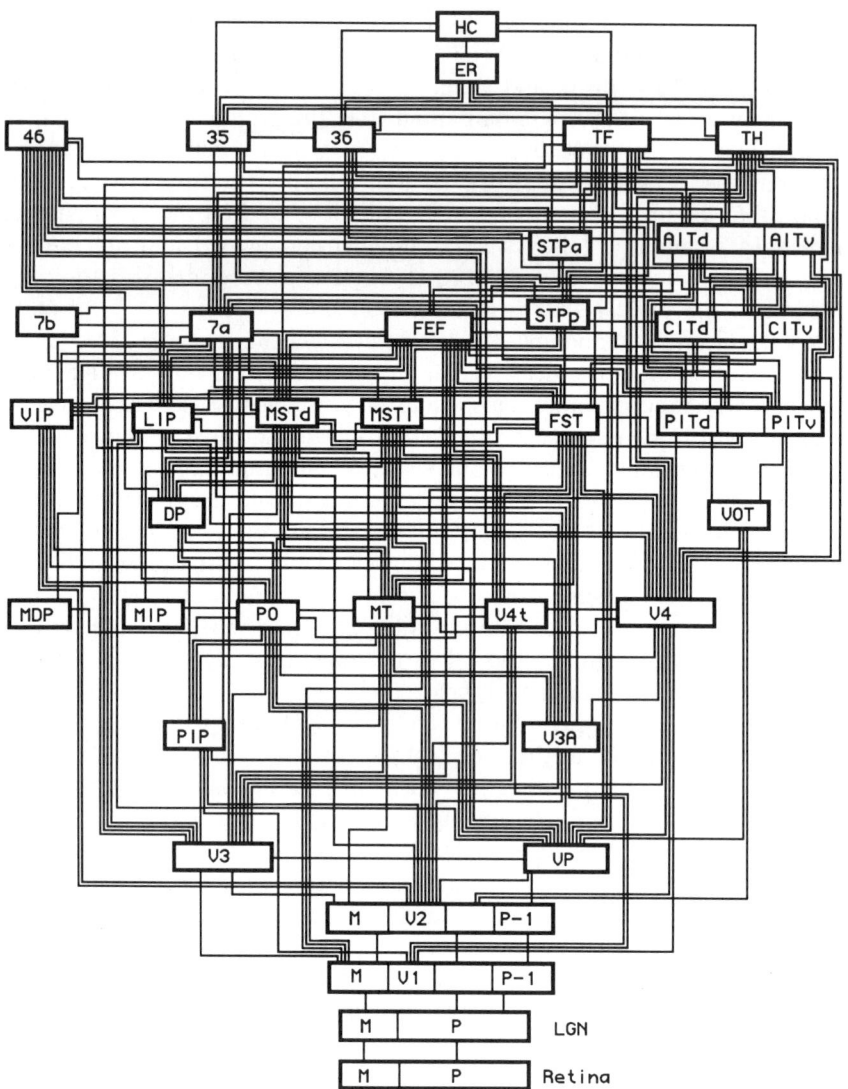

Fig. 6.1 Hierarchical view of working pattern-recognizing system: connection diagram of the visual areas of the macaque brain. The names of the boxes are largely arbitrary and may be ignored. Each line represents 10^5–10^6 connecting neurons. Many connections are bidirectional and processing is concurrent: all boxes at a given level process the entire visual field simultaneously. (After Crick's (1994) redaction of Suzuki and Amaral's modification of Felleman and Van Essen (1991), with scribal errors at each copying.)

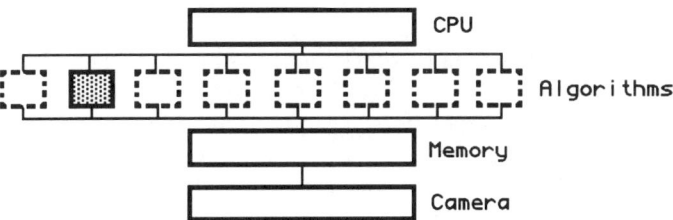

Fig. 6.2 Desk-top silicon's best current emulation of a pattern-recognition structure. Each line represents a 32-bit (or narrower) data bus running at < 100 MHz and operation is doubly sequential: one algorithm at a time and one pixel at a time. Dotted boxes are inactive while the selected algorithm is running. Current cameras mimic only the simplest qualities of the retina. Current algorithms may mimic process as high as V2 in Figure 6.1, but do not do everything that V1 does.

surfaces with the help of concepts such as good continuation, closure, similarity, proximity, and completion. Surprisingly, some of these concepts appear to be handled by the brain as early as V1 (Figure 6.1).

Despite the great difference in degree of development between carbon- and silicon-based systems, there are already certain areas in which silicon-based pattern recognition may be useful in fisheries research.

6.2 IMAGE PREPROCESSING

Raw images are seldom optimized for pattern recognition, and machine enhancement can usually improve the end results (see, for example, Figures 6.3, 6.5, and 6.10). The possibilities are numerous. Most image-analysis systems offer a variety of algorithms for each process such as contrast control, noise removal, smoothing, sharpening, edge detection, etc. There is no sharp line between the algorithms of an 'artistic' program such as Photoshop^{TM}, and the 'scientific' program of an image-analysis system. As law courts have recognized, photographs and videos are no longer trust-worthy evidence, since they can be easily manipulated to show whatever is desired (Mitchell, 1994). This imposes a certain constraint upon the user of an image-analysis system not to improve his data too much.

Examples of possible processes include some that affect the entire image, such as look-up tables (LUTs, usually pronounced lutes) which replace one pixel value with another. These are good for false color (mapping temperature to spectrum, for instance), contrast enhancement, and histogram flattening, which attempts to distribute information uniformly across the grey scale in pursuit of statistical merit.

Neighborhood transforms, which replace the central pixel of a small square by some function of its neighbors, allow both integration (smoothing and noise removal) and differentiation (sharpening and edge detection). Processing time increases as the area of the neighborhood, but almost all transforms work better in a circular area than a square. The solution is to use the largest feasible square and fill its corners with zeros. It is usually possible to limit the operation of a neighborhood transform to a selected area of the screen. The first 12-bit board-level image-analysis systems bypassed the n^2 time for neighborhood binary transforms by a clever trick: the hardware could load a 3×4 neighborhood into the memory location representing the central pixel. This could then be manipulated with a LUT in a single frame time (nominally $1/30$ s – in practice, allowing for loading the appropriate LUTs, 1 s), allowing 3×3 neighborhood operations such as shrinking, enlarging, skeletonization, etc. to be done stepwise while being watched. A 32-bit machine could do as much with a 5×5 neighborhood. It was this approach which made any sort of image analysis feasible on machines with 8 MHz clocks. Unfortunately, only limited use could be found for the 3×3 transforms which could be performed in this manner.

Segmentation – the isolation of interesting portions of the image – is often done on the basis of intensity (grey-scale thresholding), which requires that the interesting features lie in the same intensity band, and be separable from the background. With sufficient CPU speed, it is possible to compute the average intensity over an area containing both field and ground and use this value for local thresholding, but it makes better sense to strive for flat field illumination. Special methods are available for separating tangent objects in a formal manner, but the books which describe them tend to show two overlapping red blood cells in an image, where a hospital technician will scan a field of 25. Once segmented, there are various approaches to extracting further information. One is to walk the perimeter of each defined object, calculating such things as area, perimeter, caliper diameters, and centroid on-the-fly, pixel by pixel. Polygonal representation of the object can then be obtained by selecting a subset of the perimeter points, either in even steps or at extrema. The convex hull (the polygon formed by a rubber band stretched around the object) requires a fast processor but is sometimes useful.

Since pixel values are only numbers, it is possible to perform modulo-n arithmetic on them, where n is the pixel depth. More useful is the ability to perform multiimage arithmetic, such as averaging the results of many video frames to reduce camera noise. The motion of a contrasting object can be followed, producing multiple time-lapse images in a single frame, for motion studies. Uneven illumination and mottling can sometimes be reduced by subtracting a blurred image from a sharp one. Background images can be stored for later subtraction from a series of data images.

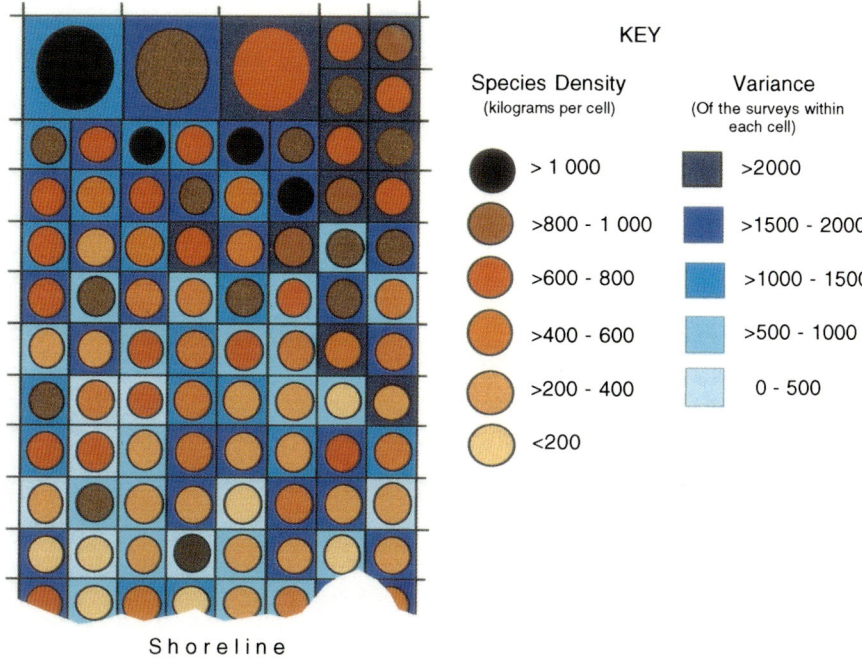

KEY

Species Density
(kilograms per cell)

Variance
(Of the surveys within each cell)

Species Density:
- > 1 000
- >800 - 1 000
- >600 - 800
- >400 - 600
- >200 - 400
- <200

Variance:
- >2000
- >1500 - 2000
- >1000 - 1500
- >500 - 1000
- 0 - 500

Shoreline

Plate 1 A means of mapping to display both biomass density and variance. Cells correspond to two levels of the spatial subdivision of a marine area. Thus it is conceived that a sea or ocean area might have a nested hierarchy of cellular data collection units (from Meaden, 1994).

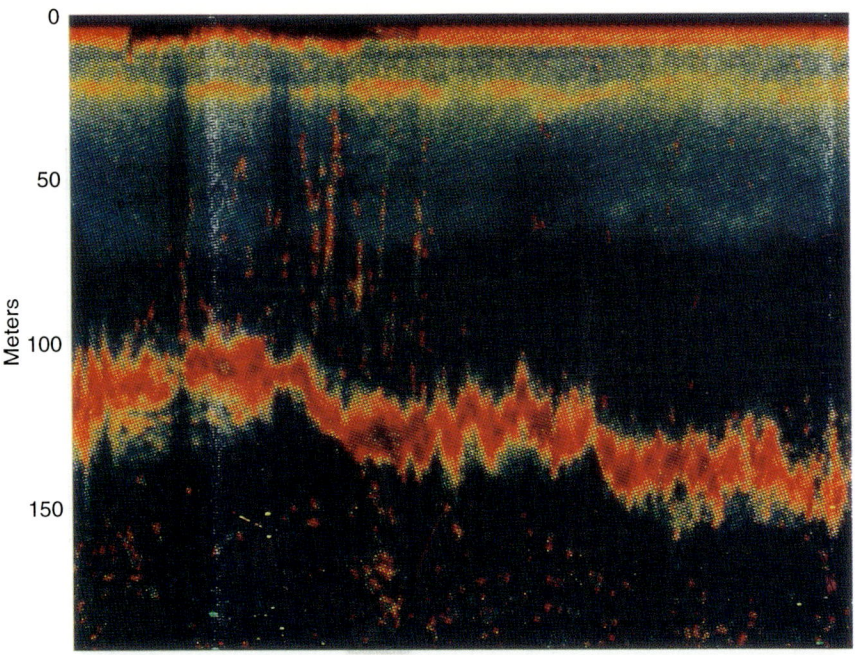

Plate 4 Graphic representation of acoustic data collected with a Simrad EK 500 (120 kHz). Net collections of fish showed that the red band between ca. 100 and 150 m was attributable to a school of mesopelagic fish. (Unpublished echogram provided by T. Knutsen, Institute of Marine Research, Norway and P. Wiebe, Woods Hole Institute of Oceanography, USA.)

Having mentioned pixel depth, some attention should be paid to cameras. The least-significant two pixels of the usual inexpensive 8 bit video camera are noise at room temperature and can be usefully discarded, freeing 2 bit planes for other uses, such as thresholding or superimposed graphics. Fourteen-bit pixel depth, with intensity resolution well beyond that of the human eye, is available (and useful for medical and astronomical images) by cooling the camera detector to liquid-nitrogen temperatures.

A final method of image preprocessing involves transforms of the entire image, of which the Fourier transform, described below, is the most generally useful.

The astute reader will notice a dearth of books entitled *Pattern recognition* in the bibliography. Those available in the libraries and bookstores of Amsterdam and Bergen tend to fall into two classes. One is abstract theoretical works with much symbolism, high goals, discussions of patterns in images, textures, grammars, and maps – and not a single image. The other is practical works containing algorithms and real images, few of which are adequate for use in the real world. We did not find a single one worth recommending. Useful books on the related but less ambitious subject of video microscopy are Inoué (1988) and Russ (1990).

6.3 IDENTIFYING PHYTOPLANKTON

Classical image analysis works by discarding information until what remains can be dealt with by the tools available. We illustrate this process by comparing the information content of several approaches to identifying a phytoplankter. Figure 6.3 shows a grey-scale image which represents what the eye has to work with, along with the outline of the binary image which is usually what a computer has to work with, most of the parameters which can be automatically measured by machine, and a taxonomic description of the organism in question.

Summarizing, the information content of the four descriptions can be ranked as shown in Table 6.1. Despite its numerical nature, such an exercise compares apples with oranges, but is nonetheless instructive. The last line of the table reveals how much information is thrown away by the abstractions which are made by contemporary algorithms. What is more difficult is to estimate the amount of information used by the retinex, which is certain to be more than the total available from any of the components of Figure 6.3. (The retinex is Land's term for the retina plus cortex, which he sees as logically inseparable.) Even for the grey-scale image, a good deal of peripheral information is necessary to determine which aspects of the picture are relevant. The total relative information used by the retinex might be many times larger than anything indicated in Table 6.1. Little of

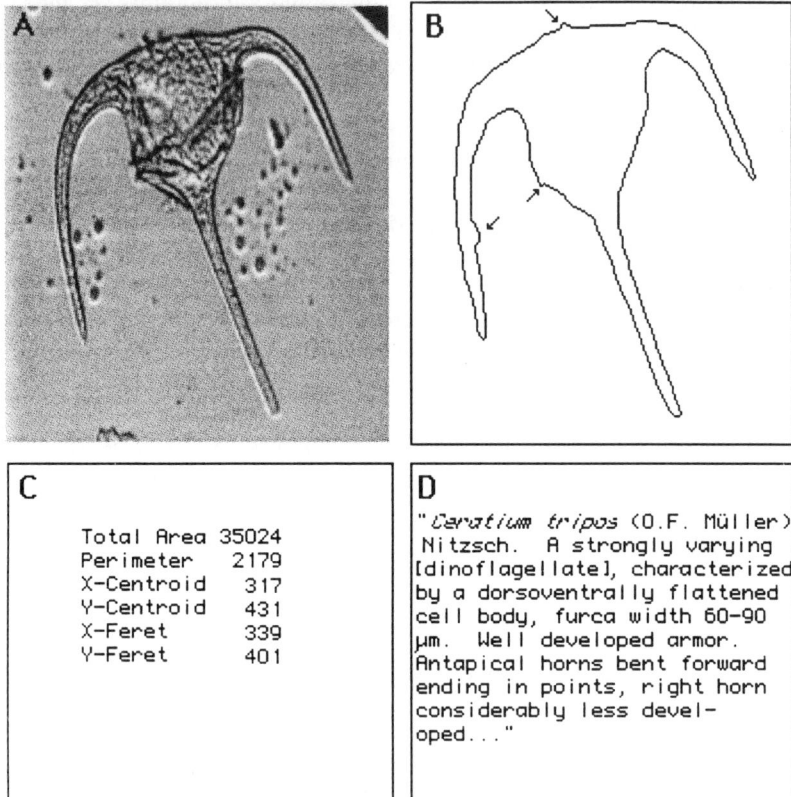

Fig. 6.3 An original grey-scale image (A) of a common marine phytoplankter, automatically reduced to its binary outline (B) and measured for some primary image-analysis parameters (C) using a Zeus[TM] image analyzer. A taxonomic description (D) of the species is translated from the original German (after Drebes, 1974). Arrows in (B) point to contour distortions arising from background material. The numbers in (C) are in pixels, but are also available from the machine in scaled form if desired.

this additional information is available to the image analyzer, which is stuck with the numbers in part (C).

The formal (Shannon) information is smaller and less reliable than the true amount for parts (C) and (D), because both of these descriptions assume additional information. In (C), one is expected to know what the various numbers signify; in (D), the words dinoflagellate, furca, armor, antapical, and horns all imply awareness of special contextual definitions. The apparent mismatch between forward and right implies knowledge of

Table 6.1 Relative information in Figure 6.3

	Grey-scale (A)	*Contour* (B)	*Image analysis* (C)	*Taxonomy* (D)
Estimated total bits	524 288	34 864	171	2160
Estimated useful bits n	130 000	34 864	1200	15 000
Comment on bit count n	75% of image is background	Image can be reconstructed exactly	Implied definitions	Implied definitions
Shannon information = ($n \ln n$)	1.5×10^6	360×10^3	8.5×10^3	144×10^3
Relative information	176	42	1	17

the standard orientation of such specimens. We have multiplied the local information by seven as an estimate of the true information conveyed by (C) and (D). Even so, the disparity between the information in the grey-scale image and the numbers available from image analysis gives an honest measure of the challenging nature of the problem.

An image-analysis (IA) system usually reports more parameters than shown in (C). Among these are such things as length, breadth, equivalent circular diameter, equivalent circular area, convexity, aspect ratio, smooth perimeter, surface area, and volume, but with the exception of a few more feret (caliper) diameters at other angles, these are derived by manipulating the few given here, and contain no new information. (There is also a small set of measures derived from local curvature, which are hard to name colloquially because the eye does not easily see what is being measured. They are also time-consuming to calculate, and at today's pixel resolution, tend to be quite jagged and noisy.)

Nevertheless, it is possible to obtain automatic separation of several phytoplankton species in mixed samples; however, be warned: this is easier to do for some species mixes than others (Estep and MacIntyre, 1989), because of the limited number of parameters available for comparison. The task is considerably less tractable if the machine is asked to separate natural communities containing many similarly shaped species and detritus, and the problem of variable shapes (from rotation of the subject or disposition of mobile body parts) remains challenging. Future improvements in pixel resolution will allow more precise shape characterization. In the ultimate IA system, pixel resolution should be comparable to that of the human eye. Printers and optimal scanners are now resolving 1200 dots per inch or about 20 μm per pixel. A 16 bit IA system at this resolution would need 2 Gbytes per image. Memories and processor speed to handle this amount of data are yet to appear at prices that grants to individual researchers can afford.

As this is written, the most productive approach to phytoplankton identification (at least in preserved material) is to contract the services of marine technicians in Eastern Europe, who will provide reliable counts of real samples at low cost.

An ambitious international research project currently underway gives an idea of what may transpire in the future. The ultimate goal is 100% (!) correct automatic classification (of diatoms) which is not dependent on size and rotation (scale and 3-D rotation invariance).

Estep, codeveloper of both *Zeus* and the de facto world-standard taxonomic software *Linnaeus II*, sees these 2 projects as the basis of a more feasible hybrid approach for the foreseeable future. *Linnaeus II* is an interactive multimedia taxonomy system which incorporates text (species descriptions, synonyms, references, etc.), pictures (from electron micrographs to satellite images), sounds and sonograms, videos of behavior, distribution maps, etc. in a smoothly functional point-and-click interface. *Linnaeus II* is distributed free to contractual ETI partners, who receive the software and CD-ROMs in exchange for taxonomic data which is added to ETI's world biodiversity database (WBD), a joint undertaking with UNESCO and others. (ETI, the expert center for taxonomic identification, is a non profit-making organization which distributes its taxonomic products at cost.) The goal is 2 million species described in a common format, to provide immediate answers to questions such as 'what anthropods and grasses are common to the colder Argentine Pampas and the Russian Steppes?' It will eventually be an on-line system accessible by e-mail, thus making available images of all described species, but it remains to be seen whether bandwidth limitations make this approach, or CDs, the method of choice for distribution of taxonomic information. There are many titles of interest to fisheries scientists.

The practical limit of the current system – not yet reached for any taxon – is about 3000 species per CD. The basic identification system is a matrix key: pick any set of observable characteristics, obtain a list of possible matches sorted by the percentage of agreement with the input set. Figure 6.4 shows the result of such a search and, since the organism in question is flexible, suggests the difficulties of asking software to make a decision about an object of this nature.

In the envisioned hybrid system – the next logical step in morphological taxonomy – the IA portion would produce images of the organism in question, and extract numerical parameters therefrom. An analysis system, something similar to *Linnaeus II* running on the same front-end machine, would access a catalog of relevant images (by CD or satellite), plus taxonomic data keyed to the numerical IA parameters. It would scan the IA parameters automatically and search the data files for matches, presenting the viewer with a graded list of the most probable species. The final pattern matching of object and stored images would then be done by the human operator, after the computers have done what they do best, namely, the time-consuming measurements and data searches. Such a hybrid system should require only modest development from currently existing programs. Interested taxonomists and image-analysis systems integrators may obtain further information from the authors.

6.4 IDENTIFYING FISH STOCKS FROM OTOLITH SHAPE DIFFERENCES

A commonly available IA operation is transformation of an image from the spatial domain to the frequency domain via a fast Fourier transform (FFT). Algorithm writers take great pride in writing an efficient FFT, because it lends itself well to clever coding, and the results not only look like black magic, but are occasionally useful.

The best way to obtain a useful grasp of what is meant by spatial and frequency domains is to make use of an IA system which performs FTs. Even if the arithmetic remains obscure, the process becomes familiar and the effect of changes can be predicted.

Humans are not without direct experience of time-to-frequency transforms, because the ear performs one in converting a time-varying pressure wave into an awareness of the individual instruments of an orchestra, or in singling out a conversion at a cocktail party. Unfortunately, possessing this ability offers little insight into how it operates.

Basically, the FT of an image is like a sonogram of a sound, in that it displays frequency and intensity. In fact, a sonogram is an FT, from the sound's time domain to the acoustic frequency domain. The frequency of an image is not that of temporal repeats, but of spatial repeats: a picket fence has a strong signal in the spacing and orientation of the pickets.

The problem is compounded by the existence of different uses of FTs. At one extreme, an entire spatial image may be subjected to a 2-D transform. The result in the frequency domain is centrally symmetric: a centered blur with indistinct structure in which frequency increases radially and direction represents orientation of repetition. If, as happens to an occasional image from a space probe, the image is contaminated with noise of a single frequency, the FT will show a bright spot at this frequency. If this bright spot is removed and the image transformed back to the spatial domain, the noise disappears. The eye uses 2-D FTs during its letter-detection process, so somewhere in Figure 6.1 is a very fast 2-D FT processor (Solomon and Pelli, 1994).

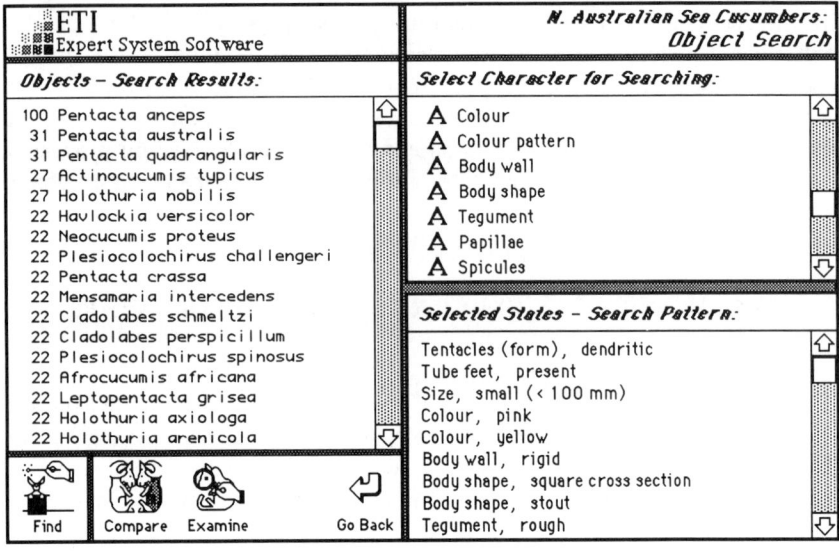

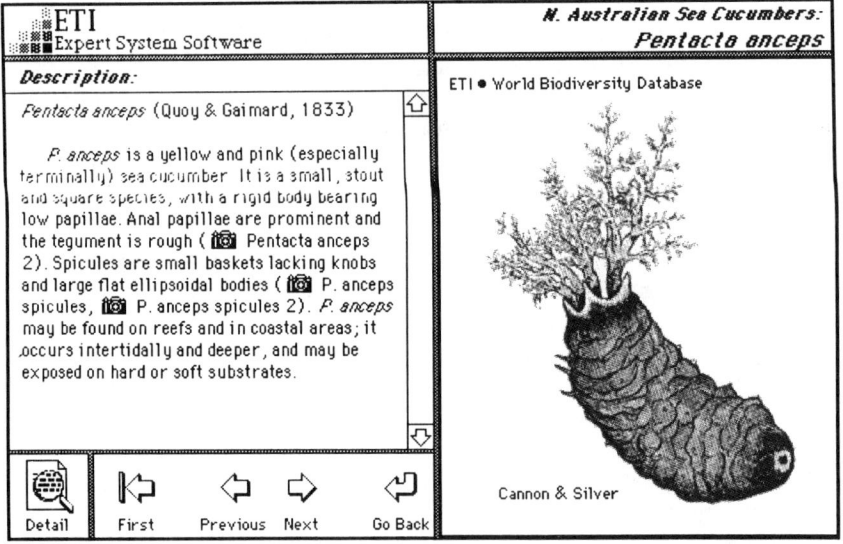

Fig. 6.4 Screen shots of *Linnaeus II*'s search module IdentifyIt™: (a) a match was requested to the test pattern at the lower right (the window needs to be scrolled to show all 22 items in the search pattern); the percentage of matching character-

However, 2-D FTs require a lot of processing power. Easier, and often more useful, is a 1-D transform of a 2-D outline. Loman (1993) describes what can be done with the outlines of organisms once they are in numerical form.

Most of us recall that any time-varying signal can be reconstructed by summing the proper assortment of pure sine waves with the correct amplitudes and phases. The FT is the device which deduces the correct amplitudes and phases. There are, however, severe limits on this process, of which the most troublesome is that the signal must be repetitive for the process to be valid. If this is not true, spurious frequencies are introduced by the edges. This is a familiar problem in the related field of time-varying signal processing: anything less than a signal of infinite duration generates spurious frequencies. Two approaches are used here. In one, the ends of the signal are artificially tapered so that amplitudes fade gradually to zero in the first and last 5% of the signal (Bloomfield, 1976). In the other, where the signal is scanned by a time window, the window itself is tapered with one of a number of competing algorithms (Beauchamp, 1987).

A similar problem arises with the edges of a 2-D image, but it does not occur with a closed contour, which repeats itself every time you go around it. One-dimensional FT analysis of an otolith outline is a textbook case of applying the right tool, and introduces no spurious frequencies.

There are many descriptions of this process in the literature, but some of them take a more complicated approach than necessary. The natural approach is to place the coordinate origin at the object's centroid, treat the (x, y) coordinates of every nth pixel around the contour as a complex number and perform a complex FT. For technical reasons, the number of points taken in the contour will usually be a power of 2. Since few contours automatically possess n^{2m} points, a common process is to use linear interpolation between existing points to arrive at the closest power of 2, perform the FT, and reverse the interpolation. The distortion introduced by this manipulation is surprisingly small, as is the time required. Figure 6.5 illustrates the process.

Figure 6.5(C) is a semilogarithmic plot of the amplitudes of the Fourier

---

istics is shown in the leftmost column; (b) clicking on a species name in the left window in the upper view will bring up the description and picture shown here. (Many other aspects of the system are not shown.) The illustrations are taken from an existing CD and are not truly relevant to the discussion of future possibilities. As the program stands, the search pattern could include IA parameters as easily as not. The more useful development discussed in the text would automatically search on IA parameters.

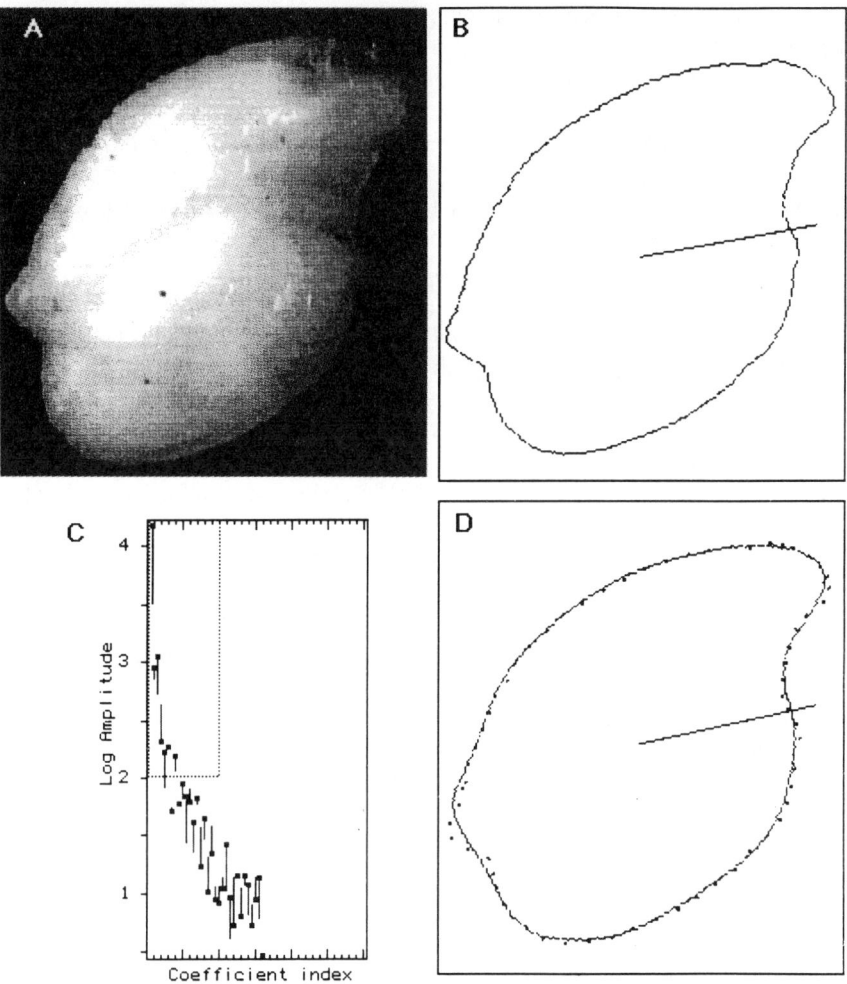

Fig. 6.5 Original grey-scale image of a fish otolith (A), which has been auto-matically reduced to its binary outline (B); FT analysis of the latter yields 16 complex coefficients (C), 7.5 of which have been reconstructed into a 2-D pre-sentation in (D). Analysis was performed using a Zeus image analyzer.

coefficients of the contour of 6.5(B). Figure 6.5(D) shows the points (some of which do not lie on the reconstructed contour) of 6.5(B) which were used in the Fourier analysis: either more or fewer points might have been chosen. The slanted line in Figures 6.5(B) and (D) is machine constructed to run from the centroid of the figure, through the selected contour point closest to the centroid (or, if desired, toward the farthest point), and out to the mean radius of the contour. In Figure 6.5(C), the vertical lines span the distance between the sine (square dot) and cosine (other end) terms of the coefficient pairs. In this example, the system initially computed 16 coefficient pairs (determined either by operator request, or by the number of contour points selected). The system allows a choice of sine and cosine coefficients, or equivalent amplitude and phase coefficients. The latter set allows trivial rotation of complex objects, while the amplitudes form a rotation-, translation-, and scale-independent descriptor of the contour.

Coefficients above the 10th or so (along the *x*-axis) are three orders of magnitude smaller than the leading term, and probably represent noise, which may be inherent in the contour or arise from digitization or interpolation. The dotted box in the upper left corner of Figure 6.5(C) was drawn by the operator, instructing the system to use only the Fourier coefficient contained therein in the reconstruction. Thus only the first seven pairs of coefficients (minus the 5th cosine term), with amplitudes larger than 10^2, were used to draw Figure 6.5(D).

Whether this is adequate resolution or not is left to the operator to decide. The projection on the left side of the otolith is not well reconstructed at this resolution, but only experience can tell whether it is needed for the purpose in hand (Estep *et al.*, 1994). The FT is an information-preserving process and if all coefficients are retained, will duplicate the original contour as exactly as digitization permits. The advantage of FTs is two-fold. The first is conceptual: the FT converts a contour from an arbitrary visual object into a mathematical object whose properties are well known and can be massaged by standard algorithms. The other advantage is data compression, arising from the possibility of discarding the high-frequency (noise) terms and using only the significant remainder for further analysis.

Further analysis treats the Fourier-coefficient data as a nearly singular rectangular matrix, as indicated graphically in Figure 6.6. The only algorithm which can deal with such difficult data reliably is the singular-value decomposition, or SVD (Golub and Van Loan, 1983). The SVD converts the data matrix **A** into a set of three matrices such that $A = USV'$, where **U** contains eigenvectors, **S** eigenvalues, and **V** rotators, which can be converted into the less comprehensible scores and loadings of any of the usual varieties of ordination or factor analysis (examples include **U**, $US^{1/2}$, or **US** for the score and **V**, $VS^{1/2}$, or **VS** for the loading, respectively) (MacIntyre, 1994). If fish breeding stocks cooperated by developing consistent differences in outline, however complex or non-obvious to the eye, such an ordination analysis of the components of an FT might allow detailed and automatic classification into different stocks.

An aspect of matrix decomposition and ordination analysis which tends

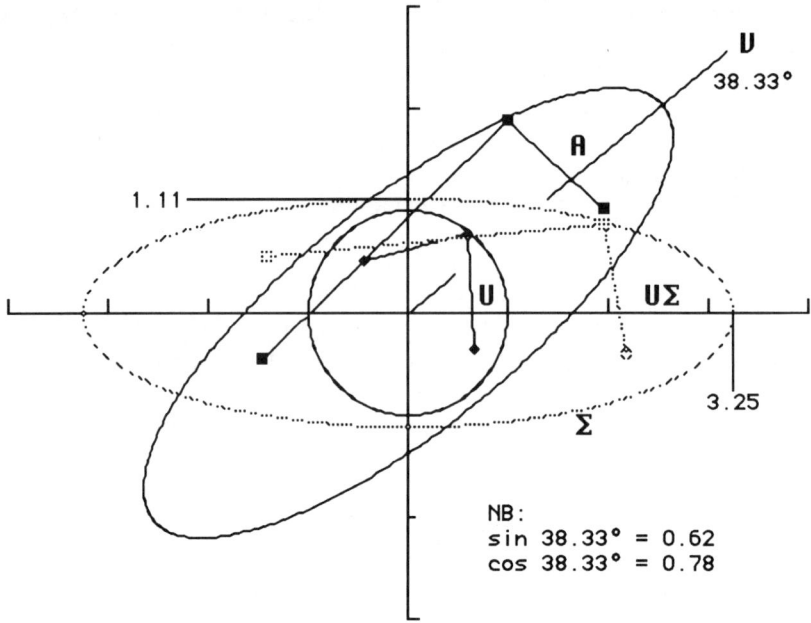

Fig. 6.6 A rank-deficient matrix rotated by the SVD, with

$$
A = \begin{bmatrix} 1.96 & 1.03 \\ 0.99 & 1.89 \\ -1.47 & -0.44 \end{bmatrix} = U\Sigma V^T = \begin{bmatrix} 0.67 & -0.37 \\ 0.60 & 0.78 \\ -0.44 & 0.51 \end{bmatrix} \begin{bmatrix} 3.25 & 0 \\ 0 & 1.11 \end{bmatrix} \begin{bmatrix} 0.78 & -0.62 \\ 0.62 & 0.78 \end{bmatrix}^T
$$

Three data points (the solid squares) lie inside an origin-centered ellipse (solid line) with major axis at angle V. Using the singular value decomposition, this ellipse can be rotated into $U\Sigma$, which lies along the axis (dotted data), and then squeezed to lie inside the unit circle, with coordinates U. This is the basis of all Cartesian methods of ordination analysis. If relative and not absolute values of the data are of interest, the matrix may be centred (and normalized) before applying the SVD.

to get lost when computers do the arithmetic is that computers do not simultaneously do the science. Transforming a matrix adds nothing to the data, and science does not enter until the results of the transformation can be given meaningful names. Ordination analysis will return its usual highly precise results for any random input.

Fortunately, we are not here interested in the meaning of the results of ordination analysis: for stock identification it would suffice if we could identify a mean contour for a species, and a set of small, consistent modifications for each breeding stock.

In any case, FTs are instructive and there will almost certainly be times when they are useful.

6.4.1 Fish scales and other contours

Nearly everything which has been said about otoliths can also be said about fish scales. The one major exception is that scales, being external, are subject to mechanical wear, which may confuse the issue. On the other hand, they are numerous, easily accessible, and – greatly appreciated by the fish – sampling is not lethal, making it easy to construct an experimental time series of scale development which can be related to environmental conditions. Fin rays and sectioned teeth offer additional opportunities for pattern analysis.

A standard example of contour FTs is the identification of airplanes for military purposes. This type of contour recognition can also be applied to the marine organisms. The immediate advantage is that FT results are independent of scaling, translation, and rotation around the optical axis. Rotation around other axes must be dealt with experimentally, as by persuading fish to swim through a narrow tube which keeps them broadside to the camera. Organisms with mobile protrusions and thus variable contours present special problems which to our knowledge have not been satisfactorily addressed.

6.5 AGE DETERMINATION AND GROWTH-RATE MEASUREMENTS OF FISH

The growth of fish is regulated by physiology as well as food abundance and quality. Specifically, daily growth cycles are a function of diurnal patterns in feeding and metabolism, while annual cycles are dependent primarily upon seasonal oscillations in food supply. At least for some fish taxa, e.g. herring *Clupea harengus* and cod *Gadus morhua*, these cycles are evidenced in otolith microstructure, i.e. daily growth in larvae and annual growth in adults are recorded as layers of calcium and organic material, which appear as rings in thin-sectioned preparations. Distance between any two rings or increments is relative to the growth of the host during the time interval corresponding to production of the two increments. This fact is extremely convenient for fisheries biologists interested in the growth dynamics of fish. To determine age, the basic if oversimplified procedure is to sit down in front of a microscope, count the increments, and convert one-to-one to the appropriate calender unit, i.e. days or years. Growth can be estimated by measuring the distances between increments. Consequently, growth rate is reflected by the ratio of interincremental distance to age (Plate 3).

The problem lies chiefly in sample preparation and subjectivity during manual selection of rings. This can be alleviated through appropriate staining and etching techniques (Secor *et al.*, 1991), which can enhance

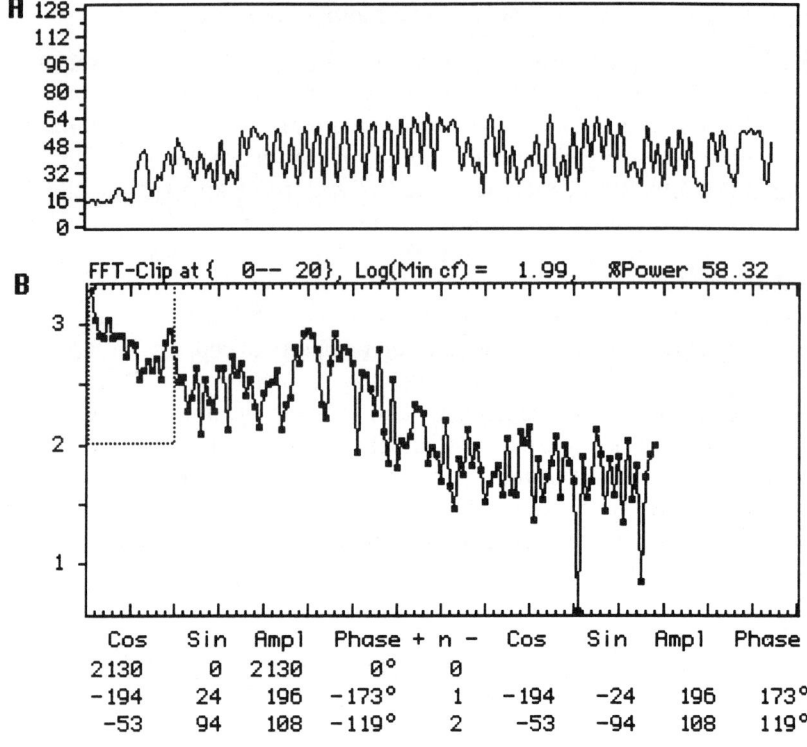

Cos	Sin	Ampl	Phase	+ n −	Cos	Sin	Ampl	Phase
2130	0	2130	0°	0				
−194	24	196	−173°	1	−194	−24	196	173°
−53	94	108	−119°	2	−53	−94	108	119°

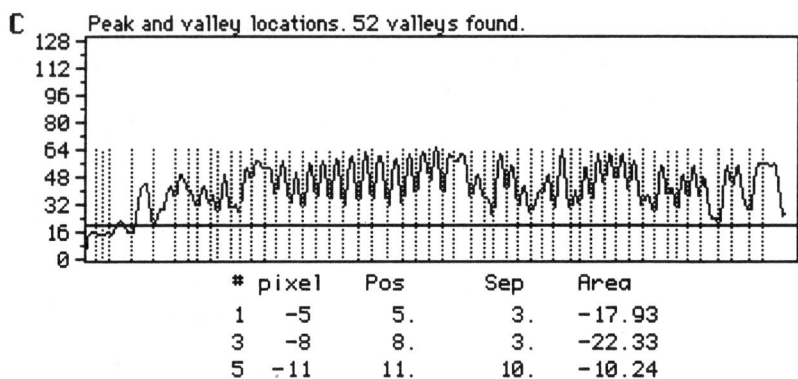

#	pixel	Pos	Sep	Area
1	−5	5.	3.	−17.93
3	−8	8.	3.	−22.33
5	−11	11.	10.	−10.24

Fig. 6.7 The same otolith shown in Plate 3 was analyzed for number of daily rings and incremental distances. Analysis was conducted along a transect (thick white line in the plate) from the center to outer edge of the otolith. Relative values for grey-level intensity (A) were converted by FFT (B) to mathematical derivations. From the inverse FFT reconstruction (C), valleys (corresponding to dark rings) were identified and labeled. The first three sets of numerical data constituting (B) and (C) are shown below the respective curves. Of particular interest to fisheries biologists

visibility of rings. Still, intercalibration exercises (Campana and Moksness, 1991) recommend stringent protocol particularly with reference to researcher subjectivity, whereby the weakest points of otolith reading were reported to include otolith preparation, proper calibration of the optics and interpretation of the image – items largely determined by the operator. A properly calibrated video/image-analysis system aids at least in reducing subjectivity and increasing accuracy and precision; effectiveness is correlated with image magnification (scanning electron microscopy (SEM) if necessary) and quality. Ironically, in poorly prepared samples and/or using suboptimal optics and microscopes, identifying increments in otoliths (just like identifying faces) may rely as much on indescribable instantaneous flashes of inspiration resulting from years of experience, as on detecting well defined alternating black and white bands. Thus for the time being otolith reading remains a pragmatic partnership between high-tech image-analysis systems and experienced laboratory assistants.

Otolith-ring analysis is another place where FTs can help. Two-dimensional (2-D) FTs can selectively emphasize contrast at the spatial frequency of the rings. The less process-intensive 1-D FT can be used to count rings in an unbiased manner. Figure 6.7 converts a radial density scan (A) of the otolith of Plate 3 into an FT whose coefficients are shown in (B). Again, the dotted box in the upper left is the operator's way of selecting which coefficients are included. In this case, it is possible to omit the first few coefficients, which has the effect of removing long slow variation and waves across the transect. The '%Power' number indicates what percentage of available information is included inside the dotted box. The table below (B) is the initial part of an exportable spread-sheet output. Both sine–cosine and amplitude–phase values of the coefficients are shown. The zero*th* coefficient is the only one which does not occur with both positive and negative frequencies, shown to the left and right of the ' $+ n -$ ' label in the heading. In (C) an inverse FT has reconstructed the density scan. A constant base line at grey-level 20 has been set; other choices are possible, such as a base line that is tangent to the deepest valleys. Peak heights and areas are measured from the baseline. The table below the diagram is again the start of an exportable spreadsheet. The machine will count either peaks or valleys, whichever is deemed appropriate.

---

are the number of valleys and intervalley distances (Sep) reported in (C). It was noted that by selecting two or more lines (thin white lines in Plate 3) transecting less noisy parts of the otolith rings, the accuracy of otolith reading could be increased; measurement of a subset of common rings along each line would be necessary in order to intercalibrate incremental distances. All analyses were performed automatically using a Zeus image analyzer and figures (A)–(C) are screenshots of reported findings.

It should be emphasized that Figure 6.7 represents a research tool whose best use is not yet known. It probably needs a way of editing peaks and valleys where operator judgement thinks editing is in order. If the single dominant frequency in (B) is selected, the output is a smooth sinewave which bridges ragged spots in the record. This might sometimes be useful. In any case, it offers new ways to think about otolith microstructure.

Pannella's (1971) recognition of daily increments in otolith microstructure is thought to be analogous to the discovery of ancient Egyptian hieroglyphs in the sense that both show consistent and repetitive patterns reflecting processes (Secor *et al.*, 1991). The hieroglyphs could not be read until the Rosetta Stone, which contained the same text in hieroglyphic, demotic, and Greek, was found and interpreted. Following Pannella's breakthrough, modern-day fisheries biologists are now challenged with constructing and deciphering their own Rosetta Stone.

An aspect of otolith analysis which seems to have received little attention is the possibility that growth is not monotonic. A similar assumption of nonmonotonicity in molluscan shells has recently been shown to be over-optimistic: 35% of fresh-water mussels of two species in a well buffered oligotrophic lake showed decreases in shell size of 10–20% over periods of 1–5 years (Downing and Downing, 1993). The only covariation noted was with location: sessile organisms in favored locations grew continuously; those in less favored locations shrank at need. The authors pointed out that the assumption of monotonicity had never before been tested in 'the long-standing literature on molluscan growth.' It is a little more challenging to examine growth monotonicity in otoliths (because it is hard to measure an otolith without discouraging further growth) but it seems that the hypothesis might now be somewhat in doubt. Checks in otoliths might represent mass loss rather than mere growth stoppage.

6.6 IDENTIFYING SPECIES FROM ECHO SOUNDERS

The possibility of using pattern recognition on sonar records of fish schools is an approach which (apparently) has not been pursued vigorously. This is at least partly because pattern recognition has been applied primarily to visual data, and sonar records are visualized only for display and then not in a convenient form for further analysis. The information available in such a display is an indication of the shape of a fish school, and intensity of return. School shape is to a certain extent species specific; intensity of return is both species and size dependent. Sonar data would be more useful – and harder to manipulate – if it were 3-D instead of 2-D (Plate 4).

A real difficulty with sonar data is that one of the dimensions is time. The sample represents a vertical look at fish in a narrow cone below the ship. The relative motion of the ship and school is seldom known. An additional dimension could be obtained with a towed array of receivers, as in seismic studies. This would provide a depth-and-distance record of fish distribution as a function of time (with less information on distribution normal to the ship track). It seems worth investigating whether some such approach would raise the cost-benefit ratio to impractically high levels: if not, it might offer a first look at the spatial organization of fish schools.

It has now been recognized that a distinct advantage of acoustic recordings potentially lies in the primary form of the data, which are target-strength signals reflected back to the source by e.g. fish. Statistical correlation of such data with specific sources is akin to shape analyses on FTs of images. On the other hand, image analyses on graphic reconstructions of acoustic data are manageable, while for most people juggling target strengths over time and depth is too abstract a concept for true insight. Contemporary research in this field therefore considers both aspects during exercises in pattern recognition of acoustic data. The application of neural networks for this work is a promising tool (Chapter 4).

6.7 NEEDED DEVELOPMENTS

6.7.1 Flat-field illumination

Attempts at computer-aided microscopy have revealed two areas in which microscope development could help appreciably.

The first is flat-field illumination. The human eye is very good at noticing out-of-focus areas in a microscope field, so microscope builders have worked hard to design lenses which deliver a field in focus out to the very edges. This is a remarkable achievement, made possibly only by computer-aided lens design. But the human eye is also very good at compensating for smoothly changing illumination, so there has been little pressure to develop substage optics which deliver uniform illumination in the object plane.

Machine analysis of an image in which circular shading is completely ignored by a human observer may show that the edges of the field are actually 40 grey-levels darker than the center. This is of no concern during visual analysis, but presents the computer with a severe problem (Figure 6.8).

The principle information which the computer has to work with (even in a color system) is the absolute intensity of illumination. It is almost trivial to explain to silicon that you are interested in grey levels between 78 and 93, and have it highlight these pixels, count them, and tell you many things about their area and shape. It is much more difficult to ask it to do

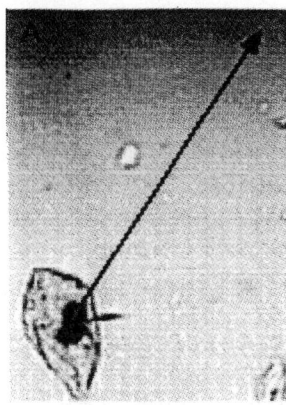

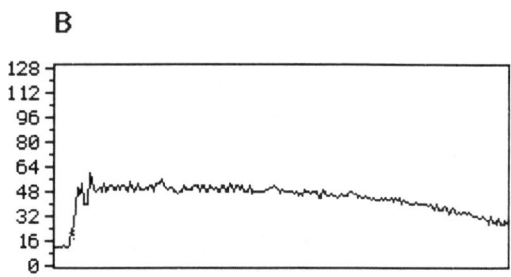

Fig. 6.8 Illumination shading (A) of marine detritus photographed with a standard charge-coupled device (CCD) video camera and research-grade microscope. Grey-level intensity (B) was measured along the line indicated in A; analysis was conducted with the Zeus image analyzer. This is a particularly good separation of object and background: in less fortunate cases, background shading occurs across the object, making it impossible to separate object from background cleanly by grey-levels alone. (Note a tendency to 8-pixel-wide vertical bands. This is not caused by microscope illumination, but is a power-supply problem on early models of the IA board employed.)

the same for areas which are 15 grey-levels different from the background around their periphery, and essentially impossible if the area in question is surrounded by objects of different intensity. The problem does not lie in finding 15 grey-level steps (over some agreed-upon distance), but in explaining the concepts of object and background, which are properties of the whole image, to a device which deals only with individual pixels.

Anyone thinking of purchasing a microscope for video microscopy should add illumination flatness to his list of specifications. Microscope manufacturers will very quickly respond to this sort of pressure.

A corollary to the need for illumination flatness is that the substage illumination path must be adjusted properly. Experience shows that this is seldom the case. Instructions for obtaining Kohler illumination (the usual desideratum) come with all serious microscopes. Nevertheless, in the course of 5 years of demonstrating image analysis with other people's microscopes, we never once found a microscope in use which was properly adjusted. This inattention results directly from the eye's indifference to shading. Consequently, the first step in a demonstration of what is possible with video microscopy is always to align the user's substage optics.

It was our intention to show a spectacular false-color picture of a shading correction field, with before and after images, to indicate what could and could not be achieved by way of machine correction of shading difficulties, since the Zeus IA system makes a noble effort in this direction. It can fit any surface up to a bicubic through an arbitrary number of points selected by the user. It thus includes circular and ellipsoidal surfaces, centered and noncentered (which are the commonest corrections needed), and also allows for superimposed nonlinear shading across the image in arbitrary directions. (Other systems use other approaches.) Each point takes the average value of a small square neighborhood. This correction function is then subtracted from background and object alike, without taking into account the transmissivity of the objects, so that the correction is at best a crude compensation. The need for such corrections introduces delays, complexity, and uncertainty, all of which are strictly unnecessary. In a properly designed microscope, there would be no need for such a clumsy rigmarole.

6.7.2 Instantaneous, calculation-free FTs

The brute-power algorithmic approach to FT works, after a fashion, but there is a more elegant alternative to obtaining frequency-domain images of microscopic samples which has been almost entirely neglected. Between the objective and ocular of every microscope lies an optical plane in which the image itself is the Fourier transform of the object. Thus, perhaps with no more than a properly placed beam splitter and a side-mounted camera, it is possible to design a video microscope which would deliver both spatial- and frequency-domain information simultaneously and instantaneously, completely bypassing the need for mathematical manipulation if all that is required is frequency-domain information.

Often, what is required is to modify the frequency domain to improve the spatial-domain image. In this case, transformation back into the spatial domains is necessary, and it is possible to imagine a reflexive system in which all transforms are performed optically, with changes made to the image in one domain appearing instantaneously in the other. The ordinary tools of a Photoshop-like program – eraser, pencil, brush, spray can, lasso, as well as more sophisticated processes – could then be used on either display and the results simultaneously observed on the other. Moving a drawing tool onto one or the other monitor would readjust the flow of information properly so that the modified view was routed to the camera and its modified transform displayed as modifications were made in the original. This (and only this) sort of immediate feedback will make FT microscopy an intuitive and useful tool for marine biologists.

We note that both CCD cameras and active-matrix displays now include nonlinear devices (transistors) at each pixel, offering a greatly improved dynamic range. In the near future it should be possible to build a transform tube such as that shown centrally in Figure 6.9, which will perform both direct and inverse FTs in a single video frame time.

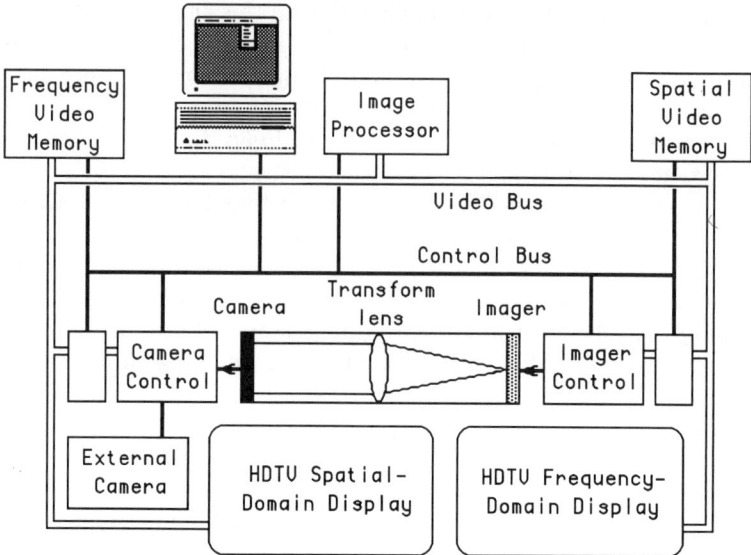

Fig. 6.9 Sketch of an image analysis system helpful for pattern recognition in both space and frequency domains. Transformation from one domain to the other is performed optically and in real time, rather than digitally after a delay. The imager might be related to the active LCD displays of contemporary notebook computers, with the same resolution as solid-state cameras. One function of the control bus is to switch the appropriate video memory to the imager control between frames.

The feedback loop between transforms implicit in Figure 6.8 might have curious properties – such as rapid degradation from pixelization (binning of area) and digitization (binning of intensity). Such problems appear not to have been studied theoretically, so there is much work to be done. There might be a need for additional circuitry next to the camera and imager controls, where empty boxes have been provided.

6.8 PROBLEMS AND PROMISES

6.8.1 Pattern recognition and dimensionality

It is suspected, based more on intuition than on any strong evidence or theory, that the ability to perform pattern recognition is related to the sort of nonintegral dimensionality popularized by coffee-table books of fractal art.

A standard search for deeper insight, beginning with the indices of, and references cited in, current books, and pursued into libraries, turned up nothing more helpful than a passing comment by Mandelbrot (1977), to the effect that a branching line in a plane might have a dimension as high as $(2 - \varepsilon)$ if it didn't intersect itself, and $(3 - \varepsilon)$ if it were a space-filling tree. Neurologists are aware that neurons are fractal (and their normal behavior chaotic), but it appears that no one is putting dimensions on neurons (Rall, 1959; Goldberger *et al.*, 1989; West and Goldberger, 1987) or trees (Nicklas, 1986).

Noting that wiring (white matter) occupies about 40% of the volume of the brain, we conjecture that there is a connectivity number for intelligent networks, similar to fractal dimension but measuring connectivity rather than space-filling, which is related to its physical complexity and to the sophistication of what it can accomplish. It is supposed that the value required for pattern recognition lies well above the value so far achieved by silicon, but may be appreciably less than the value appropriate to the mammalian brain.

Wafer-scale integration on a single silicon chip with an area of 10s of square cm is currently an active area of development, but increase of area alone is unlikely to increase the connectivity number. The Connection Machine – an array of 64 K CPU-plus-memory assemblies – allows 3-D connections in a number of topologies, but only to nearest neighbors. Something more would seem to be indicated by Figure 6.1.

We are incompetent to pursue this topic, and mention it in the hope that someone with greater insight into what we are trying to say might be able to make sense out of it. Young (1992) applied the promising-sounding 'nonmetric multidimensional scaling' to a portion of Figure 6.1, but this is only another form of ordination analysis which converts this figure into a more or less circular display with many cross-connections, but no indication of hierarchical structure.

Fig. 6.10 The Penrose tribar: what would a pattern-recognizing image-analysis system make of this?

Table 6.2 Dreyfus's classification of pattern-recognition activitites

	I *Associative*	*II* *Simple-formal*	*III* *Complex-formal*	*IV* *Nonformal*
Learning mode	learned by rote	learned by rule	learned by rule and practice	learned by perspicuous examples
Example	maze problems, trial and error	reading typed page	**recognition of complex patterns in presence of noise**	recognition of varied and distorted pattern
Program type	decision tree	*algorithm*	search-pruning heuristics	none

Italic: where we are today. **Boldface**: where we want to be.

6.8.2 Penrose

The British mathematician Penrose has a fine geometrical intuition. It was he who discovered the tribar of Figure 6.10 and the Penrose staircase that Escher used as the basis of *Waterfall* and *Ascending and Descending*, as well as a way to tile a plane with local, but not long-range, fivefold symmetry. (He is also responsible for less accessible discoveries such as twistor theory, which need not detain us here.) He explores his philosophical suspicion of strong AI – the idea that one can mimic the processes of the human brain in algorithmic silicon, and the great white hope of the pattern-recognizing community – in *The Emperor's New Mind*. His arguments range from the nonlocal paradoxes of quantum mechanics and the meaningless of simultaneity under special relativity, through a sketch of what the elusive Correct Quantum Gravity theory might be like, before drawing all of this into his conclusion.

Disappointing to those who look to computers to do difficult things, Penrose concludes that judgement – the basis of our ability to recognize patterns – is no more algorithmic than is our ability to discern beauty.

This may not mean that silicon is forever barred from recognizing the faces of its friends, but does suggest that it will not do so with designs based on the current Von Neumann (algorithmic) approach.

6.8.3 Hope for the future

In the third edition of his critique of artificial reason, Dreyfus (1993) observes that in 20 years the book has changed from a controversial position, described by a critic as short-sighted and unsympathetic, to a requiem for one of the great dreams of the last half of the 20th century. Fifty years of effort have led to a degenerating research program, one that started off with a bang with Newell and Simon and their chess-playing machines in the late 1950s, and fizzling out in the 1990s with the unheralded demise of the grandiose Japanese fifth-generation project.

Dreyfus distinguishes four classes of intelligent activities (Dreyfus, 1993). Some points relevant to pattern recognition are abstracted in the form of Table 6.2.

From this classification, the minimally useful level of pattern recognition for fisheries appears to be the ability to deal with complex-formal activities, recognizing complex patterns in the presence of noise. The state of the art is algorithm-based simple-formal analysis. True utility comes only at the nonformal level, for which there are neither programs nor ideas about how to invent them. The state of the art may reach complex-formal in the foreseeable future. Indeed, some time ago Feigenbaum and Feldman (1963) waxed enthusiastic about this possibility:

> In terms of the continuum of intelligence suggested by Armer, the computer programs we have been able to construct are still at the low end. What is important is that we continue to strike out in the direction of the milestone that represents the capabilities of human intelligence. Is there any reason to suppose that we shall never get there? None whatever. Not a single piece of evidence, no logical argument, no proof or theorem has ever been advanced which demonstrates an insurmountable hurdle along the continuum.

Dreyfus offers a sobering paraphrase of this exhortation in the form of an imagined 15th-century grant proposal from an alchemical enthusiast:

> In terms of the continuum of substances suggested by Paracelsus, the transformations we have been able to perform on baser metals are still at a low level. What is important is that we continue to strike out in the direction of the milestone, the philosopher's stone which can transform any element into any other. Is there any reason to suppose that we will never find it? None whatever. Not a single piece of evidence, no logical argument, no proof or theorem has ever been advanced which demonstrates an insurmountable hurdle along this continuum.

This somewhat pessimistic satire does not mean that we should abandon our search for computer-aided pattern recognition in fisheries. It does mean that we should not expect too much from the algorithmic approach currently implemented in desktop machines. The next generation of machines – perhaps RISC-based CPUs capable of offloading FTs to specialized signal-processing chips, and sharing their pattern-recognition with neural nets –

offers the possibility of real success with the complex-formal problems discussed above.

6.8.4 State of the art

In preparation for this chapter, we wrote to some 30 manufacturers of image-analysis and pattern-recognition equipment, describing ourselves accurately as a competitor who had gone out of business, who would be glad to give free publicity to anyone who had done anything exciting in the 3 years since we left the field. The French company Noesis, evidently the only people who felt that they had, provided an interesting insight. They had started when and as we had: two people with an idea who had developed a similar system, with similar algorithms. The only notable difference was that they had been financially successful and grown to a 50-person company, and we had not. (Were they better businessmen? Was this because they managed a cooperative agreement with both a university and the government? Of because they chose to work in France instead of Norway? Are these aspects of the same question?) What is clear is that in the 3 years of additional development they have implemented a few features that we had not (just as we had a few that they did not), and processor power has increased, making more complex analyses practical – but in principle nothing new has happened in the field. Put another way, 8 years of development seem to have exhausted the desktop possibilities of the Von Neumann approach to pattern recognition. Any 8-year-old child is better at the task.

A state-of-the-art image-analysis system is under development in the Netherlands by a commercial company and researchers from the Technical University in Delft. Like the Noesis product, this implements little that is new. It does, however, run algorithms about 50 times faster than was possible in 1990, and their $80x86$ architecture promises one more factor-of-2 speed increase. This speed increase means that algorithms such as 2-D FFTs are now feasible on desktop machines – if one can figure out what to do with them.

One of the worrying things about artificial intelligence is that for many of its theoreticians, the English language is both the medium in which they do their thinking and the object of their study. Dijkstra long ago observed that the most important asset of a programmer was mastery of his mother tongue. Yet the people who publish in computer science, and their editors, have so little grasp of English, as tool or as subject matter, that (based on analysis of abstracts) half of them cannot manage to hyphenate compound adjectives. As a group, they do not understand the subjunctive mood or the formation of latinate plurals, and cannot consistently get noun and pronoun to agree in number. This does not bode well for the future of the discipline! Nor does reading the current literature on artificial intelligence (Torasso, 1993) give one much reason to anticipate useful applications in the immediate future.

6.8.5 CPUs

We would like to suggest a radical but less Olympian view than that of philosophers such as Penrose, as to why artificial intelligence in general and pattern recognition in particular are so poorly developed. At the programming level, we suggest that the field has wasted about 50% of its time in the last 10–15 years because of the poor architectural design of CPU chips. People whose experience leads them to believe that the single-stack 8080/6800 architecture has produced state-of-the-art CPUs in the Pentium and 68040, and that development systems such as the Macintosh Programmers' Workshop (and the *x*86 equivalent) are necessary or even useful, are often so impressed by the steady increase in speed and bit width that they do not realize the inherent problems of the design.

An idea shared by Babbage, Turning, Presper, Eckert, and doubtless others, but now assigned to von Neumann, was that since data and program were both representable by binary numbers, computers could store and process them in an identical manner. A detail often hidden from the user's view is that CPUs keep their internal notes on a stack, or last-in-first-out memory structure. Overmuch attention to the data/instruction similarity lead directly to today's single-stack architecture in which data and addresses are intermixed. If you kept your personal notes in this wise, you would find yourself mingling your laundry list, business expenses, appointment calendar, experimental data, birthdays, phone numbers, and grocery list, in a single column in your notebook – an approach which no thinking person is likely to invent, let alone implement. A two-stack machine allows the programmer access to a parameter stack for his data, as in an HP calculator, while keeping its own notes about loop limits and jump addresses on the return stack.

A reviewer of this paper thought it ludicrous to suggest that poor CPU architecture and clumsy programming languages were responsible for slow progress. Because we have never had to program in C, we asked Forth inventor Charles Moore if he had a number for the performance advantage of Forth over C. His response is worth repeating.

It's hard to give you the answer you want. I can write Forth much faster than C. But your typical program? Perhaps there are four considerations:

(a) Forth is interactive. The edit, compile and test cycle is much shorter. In Forth, 10 s to correct a mistake. In C, I hear of 45 min compila-

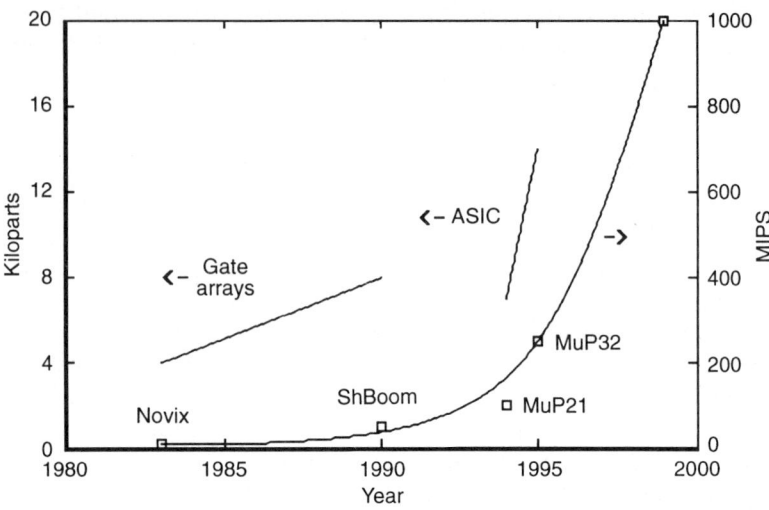

Fig. 6.11 Moore's efforts to produce Forth engines in silicon have lead to this performance graph. Kiloparts denotes thousands of transistors or gates. MIPS is an acronym for million instructions per second but really means clock frequency: the Novix executed an average of 2.8 instructions per clock cycle, the $25 MuP21, with its on-chip video processor and NTSC output, executes 5.

tions. [Others point out that C now performs incremental compilations.]

(b) Forth applications have much less code. I spec Forth code at 1–10% of C. Shorter programs are written more quickly, quite apart from measuring lines/day.

(c) Forth processors are simple. There is usually only one [hardware-determined] way to do something. Assembly programmers spend much time deciding how to best utilize on-chip resources. [Forth assembler often provides a two-to-fivefold speed increase over high-level Forth, and can be interspersed at will, either wordwise or within words. The usual approach is to code the repetitive core of a loop in assembler, leaving the setup and branch terminations in high level.] C programmers have a wealth of syntactical constructs to sort through, with cost/benefit tradeoffs to be considered.

(d) Because Forth is faster and easier to code, programmers can do a better job. This takes more time. Since I have yet to see a good C program, I conclude they can't afford to do a good job.

When Moore says that Forth processors are simple he means that while IBM and Motorola field large teams of designers to achieve the sort of performance increase shown in Figure 6.11, he works alone. A Forth compiler can be written in four lines of code, and extended by users at need. Simple does not preclude on-board address resolution and video processing, making glue chips and external graphics controllers unnecessary. Nor does simple imply reduced flexibility of the end result: Forth works everywhere from embedded controllers to planetary climate models and integrated multicomputer airport control systems handling air-conditioning, parts, food, and fuel inventories, hotel reservations, ticketing, air-traffic control, security, flight schedules and payroll. No Forth programmer we know uses a debugger more complicated than a nondestructive stack printer. Since Forth is a dictionary (rather than a linear flow of control), testing at any moment is usually limited to the most recently written definition, ideally one line of code, but under foreign operating systems, sometimes as much as 30 lines of code.

In a typical 604 page analysis of CPU architecture, Feldman and Retter (1994) mention stack machines (Koopman, 1989) on page 599, commenting favorably on their small program size, their ability to execute multiple instructions per clock period, their minimal need for accessory hardware, their efficiency at handling interrupts – and then come to the problem: 'If programs are written in other languages than Forth [everything falls apart. Thus] stack machines are not competitive players in the general computer market'. We do not often proselytize for Forth, because it appears that one's brain must be in some sense wired like Moore's to appreciate the language. The fortunate few commiserate with the many who are not so equipped. In the absence of stack machines we resort to emulating a two-stack architecture, gaining the coding advantage, but reducing performance to the machine's native level. The Harvard architecture of the PowerPC is a major step in the right direction.

6.9 SUMMARY

A rather negative view has been taken of the accomplishments of pattern recognition in fisheries, partly to offset the still prevalent impression of computer omnipotence. The human brain retains its superiority in this area: the most promising approaches seem to be those which attempt to apportion tasks properly between the machine, impossible to bore with rote processes, and the brain, clever at seeing gestalts but not very quantitative. We outline a hybrid approach which we are not in a position to follow up, but could contribute to.

The things that can usefully be done are few. However, capabilities improve continually, and there is always room for new ideas. We passed rapidly over the related field of image analysis, indicating only in the broadest terms the sort of preprocessing that can be applied to enhance features of images prior to analysis for pattern recognition, and the features which can be measured. These are well covered in books on video microscopy. Of specific interest to fisheries are identification of species (very limited success); extraction of information from otoliths and scales for stock identification, aging, and growth history (still promising areas – as they have been for a decade); and the application of pattern recognition to sonar data from fish schools (virgin territory).

REFERENCES

Anon. (1993) *PC Magazine*, 12 Oct, p. 30.

Azzopardi, P. and Cowey, A. (1993) Preferential representation of the fovea in the primary visual cortex. *Nature*, **361**, 719–21.

Beauchamp, R.G. (1987) *Transforms for Engineers*, Oxford University Press, Oxford.

Bloomfield, P (1976) *Fourier Analysis of Time Series*, John Wiley, New York.

Burr, D.C., Morrone, M.C. and Ross, J. (1994) Selective suppression of the magnocellular visual pathway during saccdic eye movements. *Nature*, **371**, 511–13.

Campana, S.E. and Moksness, E. (1991) Accuracy and precision of age and hatch date estimates from otolith microstructure examination. *ICES Journal of Marine Science*, **48**, 303–16.

Crick, F. (1994) *The Astonishing Hypothesis*, Simon and Schuster, London.

Debres, G. (1974) *Marine Phytoplankton*, Georg Thieme, Stuttgart.

Desimone, R. (1991) Face-selective cells in the temporal cortex of monkeys. *Journal of Cognitive Neuroscience*, **3**, 1–24.

Downing, W.L. and Downing, J.A. (1993) Molluscan growth and loss. *Nature*, **362**, 506.

Dreyfus, H.L. (1993) *What computers still can't do*. MIT Press, Cambridge.

Estep, K.W. and MacIntyre, F. (1989) Counting, sizing, and identification of algae using image analysis. *Sarsia*, **74**, 261–8.

Estep, K.W., Nedreaas, K.H. and MacIntyre, F. (1994) Computer image enhancement and presentation of otoliths, in *Recent Developments in Fish Otolith Research*, (eds D.H. Secor, J.M. Dean and S.E. Campana), University of South Carolina Press, Columbia, SC, 303–17.

Feigenbaum, E.A. and Feldman, J. (eds) (1963) *Computers and Thought*, McGraw-Hill, NY.

Feldman, J.M. and Retter, C.T. (1994) *Computer architecture*. McGraw-Hill, New York.

Felleman, D.J. and Van Essen, D.C. (1991) Distributed hierarchical processing in the primate cerebral cortex. *Cerebral Cortex*, **1**, 1–47.

Finkel, L.H. and Sajda, P. (1994) Constructing visual perception. *American Scientist*, **82**, 224–37.

Fredman, M.L. and Goldsmith, D.L. (1994) Three stacks. *Journal of Algorithms*, **17**, 45–70.

Goldberger, A.L., Rigney, D.R. and West, B.J. (1989) Chaos and fractals in human physiology. *Scientific American*, **262**, 42–39.

Golub, G.H. and Van Loan, C.F. (1983) *Matrix Computations*, Johns Hopkins, Baltimore, MD.

Gould, S.J. (1994) Faces are special. *The Sciences*, **34**, 36–7.

Inoue, S. (1988) *Video Microscopy*, Plenum, New York.

Koopman, P.J. (1989) *Stack computers, the new wave*, Halstead Press, New York.

Lohman, G.P. (1983) Eigenshape analysis of microfossils: a general morphometric procedure for describing change in shape. *Journal of the International Association of Mathematical Geology*, **15**, 659–72.

MacIntyre, F. (1994) The Singular-value decomposition. *Journal of Forth Applications and Research*, in press.

Mandelbrot, B. (1977) *Fractals: Form, Chance, and Dimension*, W.H. Freeman, San Francisco.

Mitchell, W.J. (1994) When is seeing believing? *Scientific American*, **270**, 68–73.

Morris, D. (1994) *The Human Animal*, BBC Books, London.

Nicklas, K.J. (1986) Computer-simulated plant evolution. *Scientific American*, **254**, 68–75.

O'Carroll, D. (1993) Feature-detecting neurons in dragonflies. *Nature*, **362**, 541–3.

Ornstein, R. and Thompson, R.F. (1984) *The Amazing Brain*, Houghton Miflin, Boston.

Pannella, G. (1971) Fish otoliths: daily growth layers and periodic patterns. *Science*, **173**, 1124–7.

Rahin, M.G. (1992) A self-learning neural-tree network for phone recognition, in *Artificial neural networks for speech and vision*, (ed. R.J. Mammone), Chapman & Hall, London, pp. 227–39.

Rall, W. (1959) Branching dendritic trees and motoneuron membrane sensitivity. *Experimental Neurology*, **1**, 491–527.

Rosky, A. and Zahn, B. (1972) Fourier descriptions for plane closed curves. *IEEE Transactions on Computers*, **C-21**, 269–81.

Russ, J.C. (1990) *Computer-Assisted Microscopy: The Measurement and Analysis of Images*, Plenum, New York.

Secor, D.H., Dean, J.M. and Laban, E.H. (1991) *Manual for otolith removal and preparation for microstructural examination*, The Bell W. Baruch Institute for Marine Biology and Coastal Research, University of South Carolina, Columbia, SC.

Solomon, J.A. and Pelli, D.G. (1994) The visual filter mediating letter identification. *Nature*, **369**, 395–7.

Srinivasnan, M.V., Zhang, S.W. and Rolfe, B. (1993) Is pattern vision in insects mediated by 'cortical' processes? *Nature*, **362**, 539–40.

Torasso, P. (1993) *Advances in Artificial Intelligence*, Lecture Notes in Artificial Intelligence No. 728, Springer-Verlag, Berlin.

Vincent, J.M., Myers, D.J. and Hutchinson, F.A. (1992) Image-feature location in multiresolution images using a hierarchy of multilayer perceptrons, in *Neural Networks for Vision, Speech, and Natural Language*, (eds R. Linggard, D.J. Myers and C. Nightingale), Chapman & Hall, London, pp. 14–24.

West, B.J. and Goldberger, A.L. (1987) Physiology in fractal dimensions. *American Scientist*, **75**, 352–65.

Wilkinson, J.H. (1978) Singular Value Decomposition – Basic aspects in *Numerical Software – Needs and Availability* (ed. D. Jacobs), Academic Press, NY.

Young, M.E. (1992) Objective analysis of the topological organization of the primate cortical visual system. *Nature*, **358**, 152–4.

Chapter seven

Computers in fisheries population dynamics

Ray Hilborn

7.1 THE BEGINNINGS

I found the first reference to computers in fisheries population dynamics in Beverton and Holt's classic 1957 book, where it states on page 309 'It has been found that an experienced computer ... can complete the calculations shown in Fig. 17.1 in about three hours'. I have used mainframe, mini, micro, personal, laptop, IBM, Amdahl, Sun, Apple, Compaq, DEC, Kapro, and quite a few other types of computers. Computers have been hopeless, worthless, fantastic, slow, big, small and super – but never **experienced**. I had to resort to the dictionary to find that in the early 1950s the definition of a computer was a person who made calculations, often with a mechanical calculator.

In the beginning, computers were people and the amount of computing that could be performed was quite limited. With reference to making dynamic predictions of the trajectory of populations over time, Beverton and Holt (1957) stated 'It is possible to obtain a generalized expression ... but this is complex and not suitable to computation'. It is clear that Beverton and Holt and the other researchers at that time were greatly limited by computational power; calculations that we routinely perform were nearly impossible for early workers. It is frightening to think what Beverton and Holt, Ricker, etc. might have done if they had had modern computational power.

This chapter reviews the history of the use of computers in the two major areas of fisheries population dynamics, estimating parameters and their uncertainty, and evaluation of alternative management policies. The emphasis is on relating the advances in computing to improved understanding of population dynamics.

Computers in Fisheries Research.
Edited by Bernard A. Megrey and Erlend Moksness.
Published in 1996 by Chapman & Hall, London. ISBN 0 412 59550 8

7.2 FITTING MODELS TO DATA USING ELECTRONIC COMPUTERS

7.2.1 Phase I: Linear fits

Modern electronic computers had a major impact on fisheries in the 1960s. Whereas computers were almost unknown at the beginning of the 1960s, by the end of this decade most population dynamists had access to mainframe computers. Things were still primitive. Most machines required computer card input and at the beginning of the decade the compilers often punched out intermediate decks of semicompiled code. Electronic terminals and disk storage of files were the exception rather than the rule. The most dreaded of all aspects of computing was the **job control language**, a series of command cards that had to precede each program telling the computer what compiler to use, what to do with the compiled program, and often even on what physical disk to put the compiled program. To those who think MS-DOS is arcane, you don't know what you missed.

The first use of electronic computers to fit fisheries models to data was primarily in the area of stock and recruitment and surplus production models. The early use of these methods relied on a linear transformation of the models so that multiple regression equations could be used to find a solution. Fitting of Gulland's method (Gulland, 1961) and Schaefer (1957) to catch per unit effort (CPUE) and catch data both relied on linear regression forms of the equations. Similarly, fitting stock and recruitment models relied on rewriting the models in a form adaptable to regression. Since single and even multiple regressions were rather easily calculated on a hand calculator, the major change electronic computers brought to these analyses was the speed at which the calculations could be done and modified. The early methods (Gulland, 1961; Fox, 1975) relied on assumptions that the population was at or near specific equilibria in order to transform the models into a regression format.

Squeezing population dynamics models into the regression packages remained the dominant method of analysis well into the 1970s. For instance Walters and Hilborn (1976) and Schnute (1977) showed how the assumption of equilibrium, used in both the Gulland and Fox methods, could be relaxed and the parameters of the Schaefer model estimated allowing for a dynamic change in abundance. However, the new methods still relied on forcing the equations into a linear regression to simplify the estimation of parameters. The advance from equilibrium assumption to dynamic fitting did not require electronic computers; the methods of Walters and Hilborn, and Schnute could still be easily done on a hand calculator.

Fitting stock recruitment models was similarly constrained by the desire to use regression equations. The Ricker (1954) model and the Beverton and

Holt (1957) models could both be transformed into linear regressions. The key problem in these linear regression forms is the assumption about error structure. The Ricker form implies a log–normal error structure, a form that is now widely accepted as the most probable (Peterman, 1981; Hilborn and Walters, 1992). However, the error structure of the Beverton–Holt equations when transformed into a linear form made little sense.

At this stage in population dynamics the emphasis was on obtaining the parameters, with less thought given to the statistical properties of the errors. Ricker (1973) was the principal voice considering the issue of noise. Further, the emphasis was almost entirely on obtaining the best parameter estimates and little consideration was given to realistically appraising the uncertainty around them.

7.2.2 Phase II: Nonlinear fitting

The second major phase involved moving beyond linear regression packages to nonlinear fitting procedures. The first authors to use nonlinear minimization in fisheries population dynamics were Pella and Tomlinson (1969), who fit the Schaefer model with an additional shape parameter to catch-and-effort data using a nonlinear least-squares procedure. Pella and Tomlinson also used what is now called the observation error approach, that is, they assumed that all errors were in the observation of the state, rather than in the dynamics of the process. Unfortunately, Pella and Tomlinson's fitting procedure remained largely ignored for a considerable time. Most attention to their paper was given to the addition of the additional shape parameter. I was unaware that Pella and Tomlinson had used the observation error method until the mid 1980s, a time when many others had 'rediscovered' this approach.

Doubleday (1976) brought fully age-structured models to life by showing that nonlinear least-squares could be used to fit such models to catch-at-age data. Up to this point virtual population analysis (VPA; Pope, 1972; Gulland, 1965) was the only dominant technique being used with catch-at-age data, and VPA was not a fitting procedure as such, but rather an algorithm to calculate apparent abundance from the data. The key distinction is that VPA provided no goodness-of-fit measure, and had no explicit or implicit error assumptions.

The potential of electronic computers and nonlinear fitting also revolutionized the analysis of length distribution data. Jones (1981, 1984), Pauly, and others (Pauly, 1987) developed the computer software ELEFAN that attempts to reconstruct the temporal distribution of recruitments and fishing mortalities from time series of catch-at-length data.

By the early 1980s, the modern method of fitting models to data had emerged. It had four basic components:

(a) a population dynamics model with parameters to be estimated;
(b) some data to fit the model to;
(c) a goodness-of-fit criterion to decide which parameters fit the data best;
(d) a nonlinear function minimization computer program to find the parameters that maximized the goodness-of-fit.

The models no longer had to be manipulated to fit into a regression package, any model could be fit. The data did not necessarily have to be a continuous time series with measurements every year, but could often include occasional observations. Most early work used simple least-squares as the goodness-of-fit, but over time maximum likelihood became the dominant approach. Finally, in the 1980s, as microcomputers started appearing on every desk, nonlinear function minimization software became quite widely available.

The generality of this approach unified many disparate assessment method. I remember Carl Walters commenting in about 1978 that Doubleday's method effectively replaced two chapters of Ricker's (1975) book. All the special cases of catch curves were part of one general catch-at-age model. This type of grand unification was also possible for other forms of analysis such as depletion methods (Hilborn and Walters, 1992, Chapter 12), and Jolly–Seber mark recapture analysis (Seber, 1982; Lebreton *et al.*, 1992).

In an effort to put this type of software on every scientist's desk, Jon Schnute spent considerable time developing a generalized nonlinear function minimization package (Schnute, 1982). Within a few years the major software makers and *Numerical Recipes* (Press *et al.*, 1986) produced nonlinear function minimization software. By the 1990s the two most popular spreadsheets EXCEL and Lotus 1-2-3 included this software.

7.2.3 Phase III: Adding additional information

The early population dynamics models required a single type of data. If you were doing stock and recruitment analysis, you wanted a time series of stock and recruitment. If you chose production models, you wanted a time series of catch-and-effort. If you were fitting age-structured models (or VPA) you wanted the catch-at-age for each year. The third phase of development involved including diverse data sources in the model fitting procedure.

Once the strait-jacket of regression packages had been escaped from, it was recognized that lots of different types of data could be used in the same model. The most widespread use of this approach has been in such methods as tuned-VPA (Laurec and Shepherd, 1983; Pope and Shepherd, 1985) or ADAPT (Gavaris, 1988) that use the basic VPA algorithm but adjust it to fit to other types of data such as effort series. Tuned-VPA nicely illustrates the basic principles of multiple data source. Traditional VPA calculates the estimated numbers-at-age and fishing mortality-at-age for each year and, as mentioned previously, is a simple recursive algorithm that involves no goodness-of-fit. However, if you assume that the fishing mortality on older fish should be proportional to fishing effort (a big if, but often assumed), then the output of VPA can be used to calculate a goodness-of-fit between fishing effort and fishing mortality on older fish. The VPA algorithm then starts with estimates of fishing mortality-at-age and uses nonlinear function minimization to find the values of fishing mortality that are consistent with the observed effort series.

Pope and Knight (1982) extended VPA to multiple species with MSVPA, calculations that required considerable computation because of the large amounts of data involved. In a parallel line of development Fournier and Archibald (1982) and Deriso *et al.* (1985) extended the age-structured model of Doubleday and allowed for additional data sources.

It is widely recognized that there are four major threads of population dynamics models: age-structured models (including yield-per-recruit and VPA), surplus-production models, stock-recruitment analysis, and length-frequency analysis. In general, the type of analysis you did depended on the type of data available. If age-specific weight, fecundity, and vulnerability data were available, yield-per-recruit was performed. Catch-at-age data led to VPA. Catch-and-effort data resulted in surplus production models; stock and recruitment data led to stock-recruitment analysis, and length-frequency data meant length-frequency analysis.

The generality of nonlinear model fitting slashed these barriers. Deriso (1980) showed that delay difference models could fully capture the age structure effects and that you could fit a model that included size and vulnerability at age (thus, yield-per-recruit), and stock and recruitment, and fit this model to catch-and-effort data, thus unifying age-structured models, surplus-production models, and stock-recruitment analysis. Schnute (1985) showed that in fact the surplus-production models were special cases of the more general models Deriso proposed. Hilborn (1990) showed that you could retain the explicit age structure and achieve the same ends as Deriso and Schnute's models.

Methot (1990) provided the ultimate unification of all the different threads in an approach he calls SYNTHESIS. At its core Methot's approach is an age-structured model. The model predicts the numbers at age, total abundance, recruitment, and the length frequency of the population. The basic parameters of the model are year class strengths, fishing and natural mortality by year and age, and selectivity to fishing or sampling gear by year and age. The SYNTHESIS model can estimate any of these parameters, depending upon what data are available. The model can use catch-at-age data, effort data, fishery independent survey data (by size, age, or total biomass), mark–recapture data, length-frequency data, growth information, etc.

The core of SYNTHESIS is a basic tautology that the number of individuals alive next year of a specific age is the number alive of that cohort the year before, less any fishing and natural mortality, less any net emigration. The size or age distribution of the fish captured by any gear (commercial, recreational, or scientific) is simply the distribution of numbers at age times a selectivity. This too is a tautology – true by the definition of selectivity. The real issues become the identifiability of parameters. The key question is how much information is contained in all of the available sources of data.

However, the world of population dynamics has not entirely jumped on the SYNTHESIS bandwagon. The Europeans continue to use tuned-VPA, the east coast of the US and Canada primarily use ADAPT, the tropics continue to use length-analysis packages. Although SYNTHESIS does, in principle, what all of these approaches do, it is not clear that it actually does it better.

Fournier *et al.* (1990) applied the generalized approach to length frequency data by applying maximum likelihood theory to this type of data. It could be argued that the approach of Fournier *et al.* is a special (and thoroughly developed) case of Methot's SYNTHESIS.

Computers not only let us estimate parameters from data, but test how well the procedures work. This is known as Monte-Carlo testing and involves simulating data sets from a hypothesized real world, and then applying our statistical procedures to these data. The advantage is, of course, that we know the 'true' answer and can thus determine the bias and precision of the estimation procedures. The first fisheries applications of this approach were done with simple models of biomass dynamics such as the Schaefer model (Hilborn, 1979; Uhler, 1979; Roff and Fairbairn, 1980), but Monte-Carlo testing is now widely practised. One of the most extensive applications was a workshop conducted by the ICES methods working group (ICES, 1988), which evaluated alternative methods for using catch-at-age data including tuned-VPA, ADAPT, and CAGEAN (Deriso *et al.*, 1985).

As the nonlinear fitting methods developed, so did the ability to understand the uncertainty in the parameter estimates. In the 1980s the method of bootstrapping (Efron, 1982) became widely practised, and many of the computer packages discussed above included bootstrap estimates of parameter uncertainty.

7.2.4 Phase IV: Using experience

Up to this point we have considered fitting population dynamics models to data: to the catch, effort, age-structure, length-structure or survey abundance data available for the stock in question. This is certainly how almost all fisheries assessments were conducted through the 1980s and many still are today. What is remarkable is that such an approach ignores every bit of experience we have ever had in fisheries – it ignores everything we have learned in the past. You simply use the data you have collected for this stock and pretend that is all you know.

This is a bit of an overstatement. The choice of which models and data to use is almost always based on experience. Further, in age-structured models the choice of the natural mortality rate was often based on experience with other stocks (Pauly, 1980). However, by and large most assessment methods have no mechanism of incorporating experience with other stocks.

There is a lot we have learned from experience that should be incorporated in our assessments. We know that some species/taxonomic groups are more prone to recruitment collapses than others. Marine mammals and sharks are easily recruitment-overfished. Clupeoid fishes have shown many collapses due to recruitment overfishing or sudden environmental change. Gadid fishes have generally proved remarkably resistant to recruitment overfishing. We know to distrust CPUE data in general and particularly when coming from fisheries showing technological development or those on aggregating species.

For many groups of fishes we can describe the frequency distribution of recruitment variation, growth rates, maturation ages, etc. Myers *et al.* (1990) for instance have collected a database on a large number of species giving estimates of several biological parameters. Databases such as this represent the collective experience of fisheries science, and this experience should be incorporated in stock assessments.

Incorporating experience is easily done and can be done in the context of traditional sum of squares, maximum likelihood, or Bayesian statistics. For instance, assume we are fitting a model M to some data D by changing some parameters P to minimize the sum of squares, SSQ. We could write the traditional fitting as the sum of squares of the model and data, given the parameters as $SSQ(M, D|P)$.

If we also had some experience E for our taxonomic group about the distribution of the values of the parameters, called $SSQ(E|P)$, then the total goodness-of-fit of the parameters to the model, data, and experience is

$$\text{total SSQ} = \lambda SSQ(E|P) + (1 - \lambda)SSQ(M, D|P)$$

where λ is the weight we assign to the experience, and $1 - \lambda$ is the weight we assign to the data. We would simply use a nonlinear search to find the values of P that minimized the total SSQ. This same approach was used in the early methods using **auxiliary data** (Deriso *et al.*, 1985).

Using past experience can also be easily done in the framework of maximum likelihood. However, past experience is most often used in Bayesian statistics, where experience is called **the prior**. In a Bayesian analysis the relative probability of the parameters is the product of its prior probability times the goodness-of-fit to the data – quite analogous to the above equation. Previous experience has been used in a number of papers; Gazey and Staley (1986) showed how prior information could be easily incorporated in mark–recapture estimates of abundance, Fried and Hilborn (1988) used preseason forecasts of salmon stock size as prior information to update in-season run forecasts, Raftery *et al.* (1988) and Givens *et al.* (1993) used estimates of virgin population size in assessing bowhead whale populations, and McAllister *et al.* (1995) used prior information on recruitment variability, acoustic survey equipment, and the spawner recruitment relationship.

Fully Bayesian analysis also provides the methods to describe the total uncertainty in the assessments, with the most fully developed Bayesian analyses of traditional assessment methods being Bergh and Butterworth (1987), Givens *et al.* (1993), and McAllister *et al.* (1995).

Environmental change is the Pandora's box of population dynamics models. Once we open the lid to environmental change, we are in trouble! Walters (1986a) pointed out that almost all models assume that parameters are either constant over time or are subject to white noise; that is, there is no systematic change in mortality, stock, and recruitment, etc. Yet increasingly, fisheries science has recognized the common occurrence of major environmental change. This means that survivals, stock recruitment parameters, etc. do change over time. Incorporating the possibility of environmental change directly into assessments is just beginning, but we all recognize it is lurking on the horizon and we are going to have to do it quite soon.

7.2.5 Summary of key issues

The major trend in the roughly 30 years since electronic computers started to be widely used in fisheries population dynamics has been increasing generality in the methods. More types of data and more types of models have been used. Other issues have included how to measure goodness-of-fit, whether to allow observation or process error (or both), what level of detail in models (or parameters) to consider, how to resolve uncertainties that remain, how to incorporate past experience, how to represent uncertainty, and how to deal with spatial structure.

We saw that the development from phase I to phase IV was in part the history of incorporating more and more data into the assessments. The early methods were able to use a consistent time series of one or two types of data. By phase III whole types of diverse data were being admitted, and in phase IV historical experience is also incorporated. To me this is the most significant development in fitting population dynamics models to data that electronic computers have made possible.

The key statistical question has always been what goodness-of-fit measure to use. The general trend has been from sum of squares to likelihood to full Bayesian analysis. A related issue is how the statistical errors are considered in the model: particularly what error distributions are assumed and are they assumed to occur only in the dynamic process, in the observation of the system, or in both. Ricker (1973) and then Walters and Ludwig (1981) and Ludwig and Walters (1981) brought this issue to the attention of most population dynamists, and Schnute (1987, 1989, 1991) has analyzed and publicized the issue in considerable detail. Most early fitting procedures assumed process error. The Pella and Tomlinson approach was the first to use an observation error estimator, which has become by far the most common approach (de La Mer, 1986; Methot, 1990; Hilborn and Walters, 1992). The advantages of observation error fitting are primarily that you do not need a consistent time series of data, and that it has been shown in Monte-Carlo testing (Punt, 1988; Polacheck *et al.*, 1993) to be more precise and less biased than process error methods. Ludwig and Walters (1981), Ludwig *et al.* (1988), and Schnute (1987) all showed how both observation and process error could be incorporated into models, and this is now gradually filtering into more and more assessments.

The various models that are used can be thought of as a wide range of special cases of the general SYNTHESIS model of Methot. The key question is what level of model complexity is appropriate. It is widely accepted that there is an appropriate match between data and models – you shouldn't have a terribly complex model if the data don't provide much information. This is why population dynamists tend to use smaller and simpler models when they have fewer data. However, there is a trap here. There are established methods of determining optimal model complexity from the data. Using either the likelihood ratio test, or the Aikake information criterion (AIC), more and more complex models can be tried to see if they provide significant improvements in the fit to the data. The accepted practice is to stop at the most complex model that still provides a significantly improved fit over its simpler alternative. This does not say that the simple models are more complex, but only that the data do not provide any evidence that the more complex model is correct.

This is clearly the wrong approach if the objective is to understand the most likely consequences of alternative management actions, or to understand the uncertainty in the dynamics of the stock. Consider a new fishery showing a declining trend in abundance. The simplest model is a linear regression, suggesting the stock will go extinct (and then negative) if fishing mortality is maintained. We would almost certainly reject the linear trend model for a population dynamics model that included some aspects of biomass dynamics. Such a model might be consistent with the population eventually coming to an equilibrium with respect to fishing mortality. However, the population dynamics model would not provide a better fit to the data than the linear trend – we would choose the population dynamics model based on knowledge of the biology of the species and experience with other fish stocks. We would not use the likelihood ratio or AIC to decide, but we would use what we know about the nature of the stock. Similarly, we might be comparing a two-parameter Schaefer model with a three-parameter Pella–Tomlinson model. For many datasets the three-parameter model would not provide a significant improvement in fit, yet we may well know enough about the biology of the species to believe that the assumption of the two-parameter Schaefer model (optimum biomass for maximum sustainable yield (MSY) occurs at one half virgin stock size) is unlikely to be true. We should not use the likelihood ratio or AIC to decide between models, but rather use the three-parameter model and include our prior knowledge about the likely distribution of optimum biomass for MSY.

Phase IV is primarily about incorporating all our experience. This is a very new field, and it is hard to tell where it will go. Smith (1994) points out how little we know of the experience of fisheries, and I suspect a major trend in the next few years will be assembling experience into forms similar to that of Myers *et al.* (1990). The statistical methods are available in Bayesian statistics, but we need to capture the history of our experience so it can be put into the assessments.

Perhaps the most intense use of electronic computers in population dynamics has been associated with determining the uncertainty in parameter estimates. In phase I the statistics of linear regression were used. In the early days of nonlinear fitting, the variance/covariance matrix of parameters derived from asymptotic theory was often presented. However, currently the three dominant methods are bootstrapping, likelihood profile, and Bayesian posteriors. All involve intense computation and could not be performed without modern computers. Bootstrapping is a very convenient and highly robust method that gained a great deal of popularity and is widely used. Unfortunately it has the disadvantage of calculating the probability of the parameter distribution given the data and can be misleading (Francis, 1992; Hilborn *et al.*, 1993). Likelihood profiles are generally easy to compute, but cannot be directly applied in decision analysis. It is now widely accepted that Bayes posteriors are the appropriate form for describing the uncertainty in our understanding of population dynamics, and methods are now available to calculate Bayes posteriors for the largest models used in fisheries population dynamics (McAllister *et al.*, 1995; Givens *et al.*, 1993).

7.3 ANALYSIS OF MANAGEMENT POLICIES

The calculation of yield-per-recruit isopleths by Beverton and Holt after World War II was probably the first large-scale use of computing in fisheries, even though it was human rather than electronic computing. Yield-per-recruit isopleths do not involve fitting models to data, but rather putting data into a model and calculating the expected consequences. This brings us to the second major use of computing in population dynamics, the evaluation of alternative management policies. As electronic computers became available it was natural that yield-per-recruit calculations should be done electronically and extended. For example, Murawski (1984) extended the single species yield-per-recruit to multispecies and multigear calculations.

Perhaps the earliest intensive use of electronic computers in fisheries population dynamics was in evaluating the consequences of harvesting policies. Ricker (1958) used hand calculators to explore the consequences of stochastic recruitment. This is the first simulation work I know of and was performed with published random number tables. As computation became available, the ability to evaluate management policies was greatly extended and numerous papers explored various aspects of optimal exploitation in a stochastic world (Larkin and Ricker, 1964; Walters, 1969; Peterman, 1975; Beddington, 1978; MacCall *et al.*, 1985).

Evaluating alternative management policies continues to be one of the major areas of computing in population dynamics. Recent work considers the consequence across the range of uncertainty in the model parameters and is usually referred to as decision theory or risk analysis (Smith *et al.*, 1993). This is really just an extension of the early work by Beverton, Holt, Ricker and Larkin, where the consequences of different actions are now simulated for a probability distribution of possible models or parameters.

The policy simulations described above explored the consequences of some very specific management policies. A technique of operations research, called dynamic programming, combined with electronic computers offered the ability to calculate the exact optimal harvest policy. Walters (1975) first applied dynamic programming to fisheries problems for a single-stock problem, with Hilborn (1976) extending the analysis to mixed-stock fisheries. The technique became popular quite rapidly and has seen widespread use (Mendelssohn, 1982, for example). The computational demands of dynamic programming are severe, and even with today's computers containing megabytes of memory, the range of problems that can be solved by dynamic programming is limited.

Evaluating alternative management policies is not particularly computer intensive as long as an assumption is made about what the decision maker knows about the population. However, when we consider the uncertainty in the estimates of parameters and its interaction with the historical management decisions, the computational problems become much, much larger. This type of policy evaluation involves evaluating both the harvest policy and the parameter estimation simultaneously. Hilborn (1979) was the first to simulate combined feedback estimation and control, using simple Schaefer models. This approach was used in two programs to evaluate feedback estimation and control in two specific management problems. The first was management of hake in Namibia and South Africa (Punt, 1988). The second, and most extensive, was a set of trials of alternative management procedures for whales performed by several groups under the general umbrella of the Scientific Committee of the International Whaling Commission (IWC) (IWC, 1988; Punt and Butterworth, 1989). In the IWC trials, the group as a whole postulated a set of possible models of whale population dynamics. Alternative models included different spatial stock structure, temporal change in model parameters, and other complications that violate the assumptions of most harvesting models. Several groups developed feedback–estimation–control algorithms for setting harvesting rules for such populations, and each algorithm was tested against the possible models that had been postulated. This detailed approach to testing estimation and harvesting algorithms is widely regarded as the ideal way to design such methods. Unfortunately, the time and expense of such a process are so demanding that so far this approach has not been extensively emulated.

A form of harvest strategies that deserves explicit mention is experimental or adaptive management policies (Walters, 1986b; Sainsbury, 1988, 1991). Experimental harvesting policies can be thought of as a special case of feedback control – one in which the management actions are deliberately designed to provide information early in the sequence of events. Sainsbury (1988) provided by far the most extensive such analysis, exploring the consequences of different spatial designs in a harvesting policy, and different lengths of experimental periods.

Finally, a special case of management policies are computed simulation games in which potential managers sit at computer consoles and manage simulated populations. Such approaches (Hilborn and Walters, 1986) are analogous to simulations used in pilot training but they have not been extensively used.

7.4 SUMMARY

Fisheries stock assessment consists of two major components: estimating the degree of belief in competing hypotheses about the dynamics of the stock, and evaluating the consequences of different management actions for the alternative hypotheses. Computers are essential to the way both of these tasks are now accomplished. One only needs to look at the scientists walking into a stock assessment meeting, most of them armed with a portable computer, to see the importance of computing. Some of the tasks now undertaken take hours or days of computer time – each new generation of computers begets more intensive computations. The laptop 486 PC I carry has as much computing power as the first IBM 360 I used in 1967 – and that IBM 360 served an entire university. This leads to two key questions: will it ever end and is it really necessary?

Will the demand for more computing power ever end? I think not. If we want to calculate the true probability of alternative hypotheses, considering all the possible data, another 10- or 100-fold increase in computer power would be most welcome. We now do many analyses that require many hours. We could try more and learn more if these tasks took a few minutes. However, increased computational power will probably not allow us to do things we do not do at present, it will only allow us to do more of what we now do.

I believe the major advance in the next decade or so will have to do with information sharing rather than raw computation. The Internet now allows us to make data available to anyone in the world. As we seek to use experience in the form of prior probabilities we are going to need easily accessible databases to draw on. The database Myers has put together on surveys and spawner recruit data (Myers *et al.*, 1990) is the first step in this development. Electronic communication also facilitates transfer of data and models and holds the hope to reduce the number of meetings and air miles logged in the process of doing stock assessment.

Is all this computing really necessary? In his 1983 stock assessment manual, John Gulland (1983) defended his failure to include computing in his stock assessment methods: 'The application of computer methods can be too easily used as a substitute for thought, or if thought is used, it is concerned with the programming procedures rather than the biological basis of the program.' This advice is as valid today as it was then. There are a number of dangers in the trend toward intense computation. Packaged methods have been the perpetual danger in fisheries computation. Someone puts together an assessment method, packages it and distributes it to a wide community. The designers know the strengths and weaknesses, but many of the users don't, and the packages are misapplied in some pretty horrible ways. This was true of the early computer packages developed on mainframe computers in the 1960s and it is true of modern packages. As we moved from mainframe computers and primitive languages accessible to a few, to desktop computers and user-friendly languages, a far greater proportion of stock assessment scientists can do their own computation and do not need to rely on fixed packages. However, since we always want to know what someone else has actually done, the advantage of the packages is that they are well understood and the widespread use of standardized methods such as tuned-VPA, ADAPT, CAGEAN, SYNTHESIS, and ELEFAN facilitates such understanding. In the hands of experienced users the pitfalls are well known. In contrast, when everyone arrives at a meeting with her or his own 'new' method, it takes a great deal of time to figure out exactly what each method is doing. Standardized packages have a major role to play and we must simply keep trying to make sure that the users are well trained in the use of the methods and understand the associated assumptions and problems.

One problem with the current focus on highly intense computation is the development of a computational elite. Butterworth once remarked that if one wanted to fully and effectively participate in the IWC Scientific Committee trials for revised management procedures, it was very helpful to be a native born English speaker and a mathematician. In the meetings where new methods are being tested and developed, the intensity of computation and statistical knowledge excludes a great proportion of stock-assessment scientists. I don't find this terribly troubling since most of the 'new work' will be discarded and the useful material will eventually emerge in an understandable and usable form. Those not participating in the intense computational debates are not missing a lot.

Gulland's view was that the basic elements of fisheries management were straightforward and more likely limited by data and understanding than computation. I have to agree. If you can measure the abundance of a species and can control the fishing effort, a sensible person cannot go far wrong in managing a fishery. Intense computation is at the very heart of modern methods for measuring uncertainty and evaluating fishery alternatives. We can do these things much better now than ever before in the past, and the main reason is computing. However, let us not forget that one of the reasons we have so much uncertainty is that we have often not measured the correct things, or that the reason we have to evaluate so many alternative policies is that we have not controlled fishing effort.

The challenge for those of us involved in computing in population dynamics is to make sure that this tool is effectively integrated into the fisheries management system so that all of our work has an impact on the conservation of our fisheries.

REFERENCES

Beddington, J.R. (1978) On the risks associated with different harvesting strategies. *Report of the International Whaling Commission*, **28**, 165–7.
Bergh, M.O. and Butterworth, D.S. (1987) Towards rational harvesting of South

African anchovy considering survey imprecision and recruitment variability. In *The Benguela and Comparable Ecosystems* (Eds Payne, A.I.L., Gulland, J.A. and K.H. Brink), *South African Journal of Marine Science*, 5, 937–51.

Beverton, R.J.H. and Holt, S.J. (1957) On the dynamics of exploited fish populations. *UK Min. Agric. Fish., Fish. Invest. Ser. 2* 19.

de la Mare, W.K. (1986) Fitting population models to time series of abundance data. *Report of the International Whaling Commission*, 36, 399–418.

Deriso, R.B. (1980) Harvesting strategies and parameter estimation for an age-structured model. *Canadian Journal of Fisheries and Aquatic Sciences*, 37, 268–82.

Deriso, R.B., Quinn II, T.J. and Neal, P.R. (1985) Catch-age analysis with auxiliary information. *Canadian Journal of Fisheries and Aquatic Sciences*, 42, 815–24.

Doubleday, W.G. (1976) A least squares approach to analyzing catch at age data. *Research Bulletin—International Commission for Northwest Atlantic Fisheries*, 12, 69–81.

Efron, B. (1982) *The jackknife, the bootstrap and other resampling plans*, SIAM-CBMS Regional Conference Series No. 38, Society for Industrial and Applied Mathematics, 1–92.

Fournier, D.A. and Archibald, C. (1982) A general theory for analyzing catch at age data. *Canadian Journal of Fisheries and Aquatic Sciences*, 39, 1195–207.

Fournier, D.A., Sibert, J.R., Majkowski, J. and Hampton, J. (1990) MULTIFAN a likelihood-based method for estimating growth parameters and age composition from multiple length frequency data sets illustrated by using data for southern bluefin tuna (*Thunnus maccoyii*). *Canadian Journal of Fisheries and Aquatic Sciences*, 47, 301–17.

Fox, W.W. (1975) Fitting the generalized stock production model by least-squares and equilibrium approximation. *Fishery Bulletin*, 73, 23–37.

Fried, S.M. and R. Hilborn (1988) Inseason forecasting of Bristol Bay, Alaska, sockeye salmon (*Oncorhynchus nerka*) abundance using Bayesian probability theory. *Canadian Journal of Fisheries and Aquatic Sciences*, 45, 850–5.

Francis, R.I.C.C. (1992) Use of risk analysis to assess fishery management strategies: a case study using orange roughy (*Hoplostethus atlanticus*) on the Chatham Rise, New Zealand. *Canadian Journal of Fisheries and Aquatic Sciences*, 49, 922–30.

Gavaris, S. (1988) *An adaptive framework for the estimation of population size*, Research Document 88/29, Canadian Atlantic Fisheries Scientific Research Committee.

Gazey, W.J. and Staley, M.J. (1986) Population estimation from mark-recapture experiments using a sequential Bayes algorithm. *Ecology*, 67, 941–51.

Givens, G.H., Raftery, A.E. and Zeh, J.E. (1993) Benefits of a Bayesian approach for synthesizing multiple sources of evidence and uncertainty linked by a deterministic model. *Report of the International Whaling Commission*, 43, 495–500.

Gulland, J.A. (1961) Fishing and the stocks of fish at Iceland. *UK Ministry of Agriculture, Fisheries and Food, Fisheries Investment* (Series 2) 23(4).

Gulland, J.A. (1965) Estimation of mortality rates. Annex to Rep. Arctic Fish. Working Group, *Int. Counc. Explor. Sea C.M. 1965(3)*.

Gulland, J.A. (1983) *Fish stock assessment: a manual of basic methods*, Wiley, Chichester, UK.

Hilborn, R. (1976) Optimal exploitation of multiple stocks by a common fishery. *Journal of the Fisheries Research Board of Canada*, 33, 1–5.

Hilborn, R. (1979) Comparison of fisheries control systems that utilize catch and effort data. *Journal of the Fisheries Research Board of Canada*, 36, 1477–89.

Hilborn, R. (1990) *Estimating the parameters of full age-structured models from catch and abundance data,* Bulletin No. 50, International North Pacific Fisheries Commission, pp. 207–13.

Hilborn, R. and Walters, C.J. (1986) Microcomputer fisheries simulations as training and teaching tools. *Environmental Software,* **1**, 156–63.

Hilborn, R. and Walters, C.J. (1992) *Quantitative fisheries stock assessment: choice, dynamics and uncertainty,* Chapman & Hall, New York.

Hilborn, R., Pikitch, E.K., McAllister, M.K. and Punt, A.E. (1993) Comment on 'Use of risk analysis to assess fishery management strategies: a case study using orange roughy (*Hoplostethus atlanticus*) on the Chatham Rise, New Zealand' by R.I.C.C. Francis. *Canadian Journal of Fisheries and Aquatic Sciences,* **50**, 1122–5.

ICES (1988) *Report on the workshop of methods of fish stock assessment.* International Council for the Exploration of the Seas, Copenhagen, Denmark.

IWC (1988) *Comprehensive assessment workshop on management,* International Whaling Commission, Report No. 38.

Jones, R. (1981) The use of length composition data in fish stock assessment (with notes on VPA and cohort analysis). *FAO Fisheries Circular No. 734.*

Jones, R. (1984) Assessing the effects of changes in exploitation pattern using length composition data (with notes on VPA and cohort analysis). *FAO Fisheries Technical Paper No. 256.*

Larkin, P.A. and Ricker, W.E. (1964) Further information on sustained yields from fluctuating environments. *Journal of the Fisheries Research Board of Canada,* **21**, 1–7.

Laurec, A. and Shepherd, J.G. (1983) On the analysis of catch and effort data. *Journal du Conseil International pour l'Exploration de la Mer,* **41**, 81–4.

Lebreton, J.-D., Burnham, K.P., Clobert, J. and Anderson, D.R. (1992) Modeling survival and testing biological hypotheses using marked animals: a unified approach with case studies. *Ecological Monographs,* **62**, 67–118.

Ludwig, D. and Walters, C.J. (1981) Measurement errors and uncertainty in parameter estimates for stock recruitment. *Canadian Journal of Fisheries and Aquatic Sciences,* **38**, 711–20.

Ludwig, D., Walters, C.J. and Cooke, J. (1988) Comparison of two models and two estimation methods for catch and effort data. *Natural Resource Modeling,* **2**, 457–98.

MacCall, A.D., Klingbeil, R.A. and Methot, R.D. (1985) Recent increased abundance and potential productivity of pacific mackerel (*Scomber japonicus*). *California Cooperative Oceanic Fisheries Investigations Reports,* **26**, 119–29.

McAllister, M.K., Pikitch, E.K., Punt, A.E. and Hilborn, R. (1995) A Bayesian approach to stock assessment and harvest decisions using the sampling/importance resampling algorithm. *Canadian Journal of Fisheries and Aquatic Sciences,* in press.

Mendelsohn, R. (1982) Discount factors and risk aversion in managing random fish populations. *Canadian Journal of Fisheries and Aquatic Sciences,* **39**, 1252–7.

Methot, R.D. (1990) Synthesis model: an adaptable framework for analysis of diverse stock assessment data. *International North Pacific Fisheries Commission Bulletin,* **50**, 259–77.

Murawski, S.A. (1984) Mixed-species yield-per-recruitment analysis accounting for technological interactions. *Canadian Journal of Fisheries and Aquatic Sciences,* **41**, 897–916.

Myers, R.A., Blanchard, W. and Thompson, K.R. (1990) Summary of north Atlantic fish recruitment 1942–1987. *Canadian Technical Report on Fisheries and Aquatic Systems 1743.*

Pauly, D. (1980) On the inter-relationships between natural mortality, growth parameters, and mean environmental temperature in 175 fish stocks. *Journal du Conseil International pour l'Exploration de la Mer*, **39**, 175–92.

Pauly, D. (1987) A review of the ELEFAN system for analysis of length-frequency data in fish and aquatic invertebrates, in *Length-based Methods in Fisheries Research*, ICLARM Conference Proceedings 13, (eds D. Pauly and G.R. Morgan), Manila, Philippines, pp. 7–34.

Pella, J.J. and Tomlinson, P.K. (1969) A generalized stock production model. *Bulletin of the Inter-American Tropical Tuna Commission*, **13**, 419–96.

Peterman, R.M. (1975) New techniques for policy evaluation in ecological systems: methodologies for a case study of Pacific salmon. *Journal of the Fisheries Research Board of Canada*, **32**, 2179–88.

Peterman, R.M. (1981) Form of random variation in salmon smolt-adult relations and its influence on production estimates. *Canadian Journal of Fisheries and Aquatic Sciences*, **38**, 1113–19.

Polacheck, T., Hilborn, R. and Punt, A.E. (1993) Fitting surplus production models: comparing methods and measuring uncertainty. *Canadian Journal of Fisheries and Aquatic Sciences*, **50**, 2597–607.

Pope, J.G. (1972) An investigation of the accuracy of virtual population analysis using cohort analysis. *Research Bulletin of International Commission for the Northwest Atlantic Fisheries*, **9**, 65–74.

Pope, J.G. and Knight, B.J. (1982) Simple models of predation in multi-age multi-species fisheries for considering the estimation of fishing mortalities and its effects, in *Multispecies Approaches to Fisheries Management Advice*, (ed. M.C. Mercer), Canadian Special Publications in Fisheries and Aquatic Sciences No. 59, pp. 64–9.

Pope, J.G. and Shepherd, J.G. (1985) A comparison of the performance of various methods for tuning VPAs using effort data. *Journal du Conseil International pour l'Exploration de la Mer*, **42**, 129–51.

Press, W.H., Flannery, B.P., Teukolsky, S.A. and Vetterling, W.T. (1986) *Numerical recipes: the art of scientific computing*, Cambridge University Press, Cambridge, UK.

Punt, A.E. (1988) *Model selection for the dynamics of southern African hake resource.* M.Sc. Thesis, Department of Applied Mathematics, University of Cape Town.

Punt, A.E. and Butterworth, D.S. (1989) Results of first stage screening trials for a proposed whale stock management procedure. *Report of the International Whaling Commission (Special Issue II)*, 191–7.

Raftery, A.E., Turet, P. and Zeh, J.E. (1988) A parametric empirical Bayes approach to interval estimation of bowhead whale (*Balaena mysticetus*) population size. *Report of the International Whaling Commission*, **38**, 377–88.

Ricker, W.E. (1954) Stock and recruitment. *Journal of the Fisheries Research Board of Canada*, **11**, 559–623.

Ricker, W.E. (1958) Maximum sustained yields from fluctuating environments and mixed stocks. *Journal of the Fisheries Research Board of Canada*, **15**, 991–1006.

Ricker, W.E. (1973) Linear regression in fishery research. *Journal of the Fisheries Research Board of Canada*, **30**, 409–34.

Ricker, W.E. (1975) Computation and interpretation of biological statistics of fish populations. *Bulletin of the Fisheries Research Board of Canada*, No. 191.

Roff, D.A. and Fairbairn, D.J. (1980) An evaluation of Gulland's method for fitting the Schaefer model. *Canadian Journal of Fisheries and Aquatic Sciences*, **37**, 1229–35.

Sainsbury, K. (1988) The ecological basis of multispecies fisheries, and management of a demersal fishery in tropical Australia, in *Fish Population Dynamics*, 2nd edn, (ed. J. Gulland), John Wiley and Sons, Chichester, pp. 349–82.

Sainsbury, K.J. (1991) Application of an experimental approach to management of a tropical multispecies fishery with highly uncertain dynamics. *ICES Marine Science Symposium*, **193**, 301–20.

Schaefer, M.B. (1957) A study of the dynamics of the fishery for yellowfin tuna in the eastern tropical Pacific Ocean. *Bulletin of the Inter-American Tropical Tuna Commission*, **2**, 245–85.

Schnute, J. (1977) Improved estimates from the Schaefer production model: theoretical considerations. *Journal of the Fisheries Research Board of Canada*, **34**, 583–603.

Schnute, J. (1982) A manual for easy nonlinear parameter estimation in fisheries research with interactive microcomputer programs. *Canadian Technical Report in Fisheries and Aquatic Science 1140*.

Schnute, J. (1985) A general theory for analysis of catch and effort data. *Canadian Journal of Fisheries and Aquatic Sciences*, **42**, 414–29.

Schnute, J. (1987) Data uncertainty, model ambiguity, and model identification. *Natural Resource Modeling*, **2**, 159–212.

Schnute, J. (1989) The influence of statistical effort on stock assessment: illustrations from Schaefer's model, in (Eds R.J. Beamish and G.A. McFarlane) *Effects of Ocean Variability on Recruitment and an Evaluation of Parameters in Stock Assessment Models*. Canadian Special Publication in Fisheries and Aquatic Sciences **108**, 101–109.

Schnute, J.T. (1991) The importance of noise in fish population models. *Fisheries Research*, **11**, 197–223.

Seber, G.A.F. (1982) *The estimation of animal abundance*, 2nd edn, Macmillan, New York.

Smith, T.D. (1994) *Scaling fisheries: the science of measuring the effects of fishing, 1855–1955*, Cambridge University Press, Cambridge UK.

Smith, S.J., Hunt, J.J. and Rivard, D. (eds) (1993) *Risk Evaluation and Biological Reference Points for Fisheries Management*. Canadian Special Publication in Fisheries and Aquatic Sciences **120**.

Uhler, R.S. (1979) Least squares regression estimates of the Schaefer production model: some Monte Carlo simulation results. *Canadian Journal of Fisheries and Aquatic Sciences*, **37**, 1284–94.

Walters, C.J. (1969) A generalized computer simulation model for fish population studies. *Transaction of the American Fisheries Society*, **98**, 505–12.

Walters, C.J. (1975) Optimal harvest strategies for salmon in relation to environmental variability and uncertain production parameters. *Journal of the Fisheries Research Board of Canada*, **32**, 1777–84.

Walters, C.J. (1986a) Nonstationarity of production relationships in exploited populations. *Canadian Journal of Fisheries and Aquatic Sciences*, **44** (Suppl. 2), 156–65.

Walters, C.J. (1986b) *Adaptive management of renewable resources*, Macmillan, New York.

Walters, C.J. and Hilborn, R. (1976) Adaptive control of fishing systems. *Journal of the Fisheries Research Board of Canada*, **33**, 145–59.

Walters, C.J. and Ludwig, D. (1981) Effects of measurement errors on the assessment of stock-recruitment relationships. *Canadian Journal of Fisheries and Aquatic Sciences*, **38**, 704–10.

Chapter eight

Multispecies modeling of fish populations

Kenneth A. Rose, Jeffrey A. Tyler, Dennis SinghDermot
and Edward S. Rutherford

8.1 INTRODUCTION

Computers played a critical role in the initial development of multispecies modeling of fish populations in the 1970s and, with the growing popularity of the individual-based approach, we believe they could again play a critical role in the near future. The importance of multispecies considerations has long been recognized in fisheries (Mercer, 1982; Daan and Sissenwine, 1989). Quantitative analysis of multispecies problems require some minimal level of digital computing power. Consequently, multispecies modeling received a great deal of attention with the initial availability of digital computing in the 1970s. However, whereas digital computing power has since increased at an exponential rate, the multispecies models used today are generally similar to those initial applications. As computing power increased, the limitation to multispecies modeling quickly became lack of knowledge about biological interactions. Thus, while multispecies modeling shared the initial burst of activity with digital computing, it did not continue to advance as computing power has. We argue that with the growing interest in the individual-based modeling approach, multispecies modeling will experience another burst of research activity. Computing will again become a limiting factor in analyses because the individual-based approach is very computing intensive.

This chapter expands on the role of computers in multispecies modeling of fish populations. First, the importance of multispecies considerations in fisheries is discussed. Second, the history of multispecies modeling of fisheries, and its relationship (or lack of relationship) to increases in computing power is briefly reviewed. Third, an introduction to individual-based

Computers in Fisheries Research.
Edited by Bernard A. Megrey and Erlend Moksness.
Published in 1996 by Chapman & Hall, London. ISBN 0 412 59550 8

modeling is provided. This is followed by a description and analysis of an individual-based model of yellow perch and walleye in Oneida Lake, NY. The analysis demonstrates the individual-based approach applied to a multispecies situation, the importance of interspecific considerations for understanding population responses, and the critical role computing power could once again play in multispecies analyses. Finally, a summary and a discussion of future directions is given.

8.2 MULTISPECIES MODELING IN FISHERIES

8.2.1 Multispecies interactions

Interactions among fish species in a system can be biological or technical. Biological interactions include competition for a common resource (e.g. prey, or space) and predator–prey relationships. Technical interactions occur when more than one species is caught by the same fishery. While the difficulties of quantifying biological interactions are well known, technical interactions can also be quite complex in some situations (Laurec *et al.*, 1991).

The importance of interspecific interactions to fish population dynamics and fisheries management is well recognized (Mercer, 1982; Hennemuth, 1979; Gulland, 1989; Daan and Sissenwine, 1989; Gulland and Garcia, 1984; Kerr and Ryder, 1989). Few fish species live in complete isolation from other species, although the degree of interaction among species can vary greatly. Several major workshops with a focus on marine systems have been held on multispecies approaches to fisheries management (Daan and Sissenwine, 1989; Turgeon, 1982). Examples of the importance of multispecies fish interactions in freshwater systems include the ideas underlying top-down versus bottom-up food web manipulations (Kitchell, 1992; Crowder, 1988), the effects of species invasions on Great Lakes fish communities (Kitchell and Crowder, 1986; Regier and Hartman, 1973), and the issues involving stocking of shad to enhance game fish production in many North American lakes and reservoirs (DeVries and Stein, 1990). Bax (1991) applied a modeling approach to six marine systems and estimated that > 50%, and for some systems > 90%, of the fish biomass loss was due to fish consumption.

8.2.2 Multispecies modeling

Definitions

For the purposes of this chapter, multispecies models are considered to be those that include the dynamics or energy flows of two or more fish species. Other aspects of multispecies analyses that focus on population estimation and statistical methods, or focus on one fish species while treating others as forcing functions (e.g. as terms in a spawner–recruit function), are not discussed. This review is not intended to be comprehensive. A classification scheme is used to organize this review, while recognizing that any classification scheme is somewhat arbitrary. Results related to fishes are emphasized, even though fish were not the major focus for some of the models.

Perhaps one of the first multispecies models was developed in 1928 by Volterra, who presented a series of simple coupled single-species models of different interactions (e.g. competition, predation, self-regulation). Much of the paper is devoted to defining conditions under which approximations hold allowing for an analytical solution. Multispecies modeling, as well as ecological and ecosystem modeling in general, became more feasible (especially for nonmathematicians) with the advent and availability of digital computing in the late 1960s and during the 1970s (Patten, 1969; Riffenburgh, 1969; Paulik, 1969; Hackney and Minns, 1974), although some still promoted analog computing (Silliman, 1969). The necessity for mathematical elegance was replaced with brute-force solution via numerical simulation. There was an initial pulse of activity and enthusiasm during this period. Most subsequent multispecies modeling efforts have used similar approaches to those used during the pulse of activity in the 1970s period, and have focused on well-studied systems because of the large data requirements for multispecies analyses. Usually, these systems were exploited by fisheries which provided the rationale for data collection and analysis.

For purposes of discussion, multispecies models can be classified as budget models (static or time-varying), coupled single-species models, and holistic models. Budget models are typically representations of biomass or energy flows among the compartments of a food web and can be static or temporally dynamic. Budget models have been used to identify the major source and sinks for biomass and to derive higher-order indices such as ascendancy and system throughput (Ulanowicz, 1986). Coupled single-species models and holistic models differ from budget models by simulating the processes that comprise the net flows among compartments. Coupled single-species and holistic models are usually time-varying, and differ in their complexity and representation of the interactions among species. Coupled single-species models tend to focus on the dynamics of a few species in the system and to treat the interaction between them in a simplified manner with a lumped parameter approach. Holistic models tend to include more compartments of the ecosystem and to treat the interaction among species in more biological detail, rather than as a single parameter.

Budget models

Budget models have been applied to a variety of marine systems. Georges Bank has been represented with about 10 compartments ranging from bacteria to two/three-fish compartments (Cohen *et al.*, 1982; Fogarty *et al.*, 1987; Sissenwine, 1984; Sissenwine *et al.*, 1984). These models estimated that fish consume 60–90% of their own production, and that the consumption of prerecruits of one fish species, silver hake, exceeded the total productivity of the exploited finfish and squid community. Walsh (1981) compared 13-compartment (two/three-fish compartments) carbon flow budgets of the Peruvian coastal ecosystem before and after overfishing of anchovy. He concluded that the decline in anchovy grazing pressure as a result of overfishing apparently led to increased plankton, sardine, and hake standing stocks, and ultimately caused a decline in water concentrations of oxygen and nitrate. Jarre *et al.* (1991) also presented a budget model of the Peruvian upwelling system. They included eight fish compartments and compared budgets and system-level indices among three periods (moderate anchovy abundance/limited fishery, high anchovy abundance and high fishery efforts and catches, post-fishery collapse). They discussed the changing role of anchovy in the budget and how its role as prey for top fish predators was replaced by other fish species including some anchovy predators. In a series of three papers, a 12-compartment (four-fish compartments) budget of biomass flows of a coral reef ecosystem in the Hawaiian Archipelago was constructed and analyzed (Polovina, 1984; Atkinson and Grigg, 1984; Grigg *et al.*, 1984). One of their conclusions was that coral reef communities are top-down regulated by predation (seabirds, monk seals, and fishes), rather than nutrient limited. Pace *et al.* (1984) constructed a 17-compartment (two-fish compartments) energy flux budget of a generic continental shelf food web distinguishing among trophic, dissolved, detrital, and respiratory flows. They also expressed the budget as a system of differential equations, with an equation describing the time rate of change of each compartment. Specifying within-year variation in the driving variable of water nitrogen permitted simulation of time-dependent dynamics. Different regimes of nitrogen input affected simulated fish production, with an upwelling-like regime (gradual changes) resulting in about a 10-fold increase in fish production compared to the other regimes that involved nitrogen pulses.

Coupled single-species models

A conceptually simple approach to multispecies modeling is to couple single species models with a term that represents the effects of one species on another. Such coupled single-species models tend to be heuristic because of the over-simplified representation of species interactions. May *et al.* (1979) used a series of Lotka–Volterra-like, predator–prey models to examine interactions among various combinations of krill, cephalopods, baleen whales, and sperm whales and harvesting. They concluded that maximum sustainable yield (MSY) was useful for top predators but not for other populations in the ecosystem, that the inherent time scales of different populations need to be recognized for effective management, and that harvested populations were less resilient to perturbations than virgin (unexploited) populations. Allen and McGlade (1986) combined a multispecies fish model based on a Lotka–Volterra representation with a dynamic model of fishermen and embedded both in a spatially explicit grid of cells. Their emphasis was on nonequilibrium analysis and the importance of not only modeling the fish but also the behavioral aspects of the harvesters.

Coupled logistic population models have also been used to examine simple multispecies fisheries (Pope, 1976; Allen and McGlade, 1986; Strobele and Wacker, 1991; Brander and Mohn, 1991). Pope (1976) concluded that the total yield from an interactive two-species system would be less than the sum of MSYs of each species. Brander and Mohn (1991) questioned this conclusion and showed that whether the total MSY was less or greater than the sum of individual species' MSYs depended on how predation mortality on prey species was represented under single-species calculations. Strobele and Wacker (1991) offered a variety of extensions to simple two-species linked logistic models by expressing interaction terms for the conditions of predation, competition, and mutualism, and by expressing harvesting as removal of one species only or of both species. They concluded that, depending on the interaction term and species being harvested, complex, even nonconvex, yield (sustainable harvest) versus effort curves can be generated. Yodiz (1994) eloquently captured the fundamental drawback to the coupled single-species models approach to multispecies analyses by showing that the choice of the functional form of the interaction terms can greatly influence model predictions.

Age/size structure has been added to single-species linked models. Murawski (1984) proposed a multi-species, mixed fishery model for situations when technical interactions are dominant and biological interactions are minimal. The model was based on age-structured, single population models and used to compute equilibrium values of yield per recruit. The model was applied to the four-species, six-fishery bottom trawl fishery on Georges Bank (Murawski, 1984) and to the 15-species, six-fishery demersal community of the Gulf of Maine (Murawski *et al.*, 1991). In the latter example, discards were important in both small and large mesh fisheries implying that, even if small-mesh fishing was curtailed, the benefits to long-lived species yields would be marginal because discard would simply occur in the large mesh fisheries. Brander (1988) used a similar equilibrium yield-per-recruit analysis to examine interactions between Norway lobster and its dominant predator cod in the Irish Sea where they each appear as bycatch in the other's trawl fishery. Wilson *et al.* (1991a) used an age-structured, five-species model where first-year survivorship was determined from spawner–recruit relationships and subsequent survival determined by natural and fishing mortality rates. Growth was based on a von Bertalanffy formulation. The five species interacted via a community predation function that lowered first-year survival of each species in a predetermined order as total biomass approached a specified carrying capacity. Model results indicated that only when individual species in the system were in a chaotic state did the model qualitatively replicate the dynamics observed in real multispecies systems, and that total biomass showed much less temporal variation than its component species. Wilson *et al.* (1991b) added a fishery simulator to the five-species model which assumed fishermen were profit maximizers who can allocate their effort according to the price and abundance of each species. Fishing effort caused higher variability in species abundances when total biomass was able to attain the community carrying capacity, thereby activating density-dependent effects that desta-bilized individual species.

Examples of age/size-structured, linked single-species models of fresh-water systems include Strange *et al.* (1993) and Sharov and Kriksunov (1991). Strange *et al.* (1993) used coupled Leslie models to examine the relative importance of density-dependent and density-independent factors affecting brown trout, Tahoe sucker, and rainbow trout interactions in a California stream. The model was formulated such that first-year survivor-ship was affected by stream discharge, growth and fecundity of rainbow trout were reduced in the presence of brown trout, and brown trout pre-dation on Tahoe sucker depended on brown trout and sucker abundances. Variation of stream discharges under two different initial configurations of the three-species community showed that effects of density-independent factors on population dynamics can differ depending on the biological state of the community. Sharov and Kriksunov (1991) used a similar age-struc-tured modeling approach to examine the competition for food during the first year of life between vendace and lake smelt in Pskov-Chud Lake. Changes in the mortality rates of first-year cohorts and of adults of one species affected the dynamics of the other species.

Holistic models

Many holistic models, by definition, could be included here because they include more than one fish compartment. We simply highlight a few examples. Laevastu and his colleagues have formulated and applied multispecies biomass-based models (PROBUB and DYNUMES are spatially-explicit; SKEBUB is a one-box version) to a variety of marine systems including the Bering Sea, western Gulf of Alaska, Balsfjord (North Norway), and Georges Bank (Bax, 1985; Bax and Eliassen, 1990; Laevastu and Larkins, 1981; Laevestu *et al.*, 1982). Patten's (1975) linear model of Lake Texoma is an early example of an holistic model, with six-fish compartments among the 33 total compartments. Ploskey and Jenkins (1982) simulated the biomasses of six fish types defined by their source of food (e.g. plants, benthos, detritus), treating YOY separately from older fish. They applied the model to DeGray Reservoir in Arkansas and concluded that reservoir fishes are efficient grazers that can overrun their food supply and that addition of prey fish or size limits on sport fish would not provide long-term benefits to piscivores. A 48-compartment general ecosystem model that simulates water quality, algae, zooplankton, benthos and four fish groups (each divided into five life stages) has been applied to Cayuga Lake, NY (Tetra Tech, 1980). Clupeids fish group biomass was predicted to increase under scenarios of increased numbers of power plants because nutrient-rich bottom waters used for power plant cooling were discharged to surface waters where they increased primary and secondary production. Predicted biomass of the other fish groups was unaffected by the number of power plants because the other direct effects of power plant operations (entrainment, impingement, and thermal and toxic discharges) were minimal.

Perhaps the most famous holistic fisheries model was that proposed by Anderson and Ursin (1977), which also was the basis of the multispecies virtual population analysis (MSVPA) approach. Their model included the dynamics of nutrients, phytoplankton, zooplankton, detritus, and age-structured Beverton–Holt-like models of the average weight and numbers of individual fish species, with the full model expressed as a system of 308 ordinary differential equations. Consumption by animals was based on predator to prey size ratios. They applied a 21-variable version (11 fish species) to the North Sea system and estimated the state of the virgin stocks (i.e. no fishing), noting that cod replaced man as the dominant predator, and that maximum possible yield, under suspiciously high effort in their opinion, was about twice the actual 1970 yield. MSVPA can be viewed as a simplification of the Anderson and Ursin model (Sparre, 1991). MSVPA has been used extensively in multispecies analyses of exploited marine systems, forming the cornerstone of multispecies analysis performed by the Multispecies Assessment Working Group of ICES (Anon., 1990; Pope, 1989).

General results

As evidenced by the above review, multispecies modeling methods and the systems analyzed are diverse, making generalization difficult. However, four general results emerged from the review:

(a) enormous amounts of biological information are required for model formulation and corroboration;
(b) interspecific interactions are important;
(c) abundance and biomass at the community level are less variable than on an individual species level;
(d) the fact that methods have not changed much over time, while computing power has increased enormously, suggests computing power is not limiting.

Practically all of the papers reviewed mention the enormous amounts of data and information required for configuring a multispecies model. Data limitations on multispecies modeling become especially critical when such analyses are used for site-specific assessments and management (Mercer, 1982; Grosslein *et al.*, 1983; Patten *et al.*, 1983; Haar *et al.*, 1981; Brugge and Holden, 1991; Murawski, 1991). The concept of competition and how to quantify it also remains a contentious issue in community ecology (Kerr and Ryder, 1989; Fausch, 1988).

Biological and technical interactions tend to be important, making multispecies systems tightly coupled. Effects on one species affect other species, implying caution when interpreting results from a single-species analysis that typically ignores, or treats as constant, nonfishing losses. Whether multispecies interactions are universally important in marine and freshwater systems is unclear. Our sample is probably biased because multispecies models are typically applied to situations where interspecific interactions are thought to be important. An important aspect of predicting fish population dynamics is determining when a single-species analysis would provide an adequate approximation and when multispecies interactions must be explicitly represented.

Total fish abundance or biomass tends to exhibit lower interannual variability than the abundances or biomasses of the individual species. This result has been reported based on empirical data (e.g. Sutcliffe *et al.*, 1977) and modeling results (e.g. Wilson *et al.*, 1991a). Thus, while the dynamics of any individual species may appear highly variable and even chaotic, the multispecies assemblage tends to display more predictable behavior.

Computing capabilities do not presently constrain multispecies modeling. We interpret the lack of discussion of computing speed and precision in most of the multispecies modeling papers reviewed as an indication that computing was not limiting or infringing on the development or running of the models. With few exceptions, the types of models being developed and applied today are similar to the early models. Few, if any, strain the computing resources found on today's desktop computers. To illustrate this, remember that Anderson and Ursin (1977) felt it necessary to include a section entitled 'A reduced model for rapid computation' and to state that each year of simulation (of their North Sea model) required 3 min of CPU time on a 'fairly fast' IBM 360/165 computer. To put this into perspective, today's workstations offer 100–1000 times the computing speed of the IBM 360/165. This situation may be changing with the rise in popularity of individual-based modeling. Individual-based modeling is computing intensive and can exceed the capabilities of today's fastest workstations.

8.3 INDIVIDUAL-BASED APPROACH

The individual-based approach to modeling is gaining popularity in ecology in general (DeAngelis and Gross, 1992; DeAngelis *et al.*, 1994; Huston *et al.*, 1988) and specifically with fish (Van Winkle *et al.*, 1993; Persson and Diehl, 1990; Crowder *et al.*, 1992). The basic appeal of the individual-based approach is that it is conceptually straightforward and simple. Individuals are the integrating unit in nature. Almost all data collected on fish are based on measurements of individuals. Accurate modeling of fish requires following multiple traits (e.g. length, weight, age) simultaneously, some of which covary together. Fish also exhibit great plasticity in growth and other processes as well as large variation in recruitment, which cause great consternation for developers and users of models. In theory, models based on simulating many individuals should be easier to formulate because many of the data are individual-based, and should be capable of generating realistic predictions because multiple traits can be followed simultaneously. Further, the individual-based approach also lends itself to spatially explicit analyses because it is easier to simulate movement when individuals are represented than when the population is treated as an ensemble. In an ideal situation, if all individuals in a population or community are represented adequately, then the population prediction is simply the sum of individuals.

A common simplification in individual-based models of fish is to simulate individuals for a restricted portion of their life cycle (usually one or more early life stages). The simplest models deal with individuals of a single cohort for a small portion of the life cycle in a spatially homogenous environment (Beyer and Laurence, 1980; Adams and DeAngelis, 1987; Rice *et al.*, 1993). Spatial heterogeneity has been represented explicitly by coupling an individual-based model of early life stages with a hydro-dynamics model (Walters *et al.*, 1992; Hinckley *et al.*, in prep.), but these models tend to be biologically simple (Tyler and Rose, 1994). The addition of reproduction and development through early life stages of the first year of life results in model predictions of YOY dynamics, either for one box (Rose and Cowan, 1993; DeAngelis *et al.*, 1991) or for spatially explicit environments (Jager *et al.*, 1993). Some of these early life stage, individual-based models have also been extended to include toxicant bioaccumulation (Madenjian and Carpenter, 1993) and toxicant effects (Rose *et al.*, 1993a).

We have been involved in a project that has attempted to simulate the entire life cycle of fish by following the progeny of surviving YOY for multiple years as they mature and spawn. These full life cycle models either follow individuals throughout their life time (e.g. Rose *et al.*, in prep-b) or couple an individual-based model of early life stages with an age or size-structured adult cohort model (Rose *et al.*, in press). We are beginning to apply models to the simplest multispecies situation of two interacting species (Rose *et al.*, in prep-a; Clark and Rose, in press); the yellow perch–walleye model of Rose *et al.* (in prep-a), is used as an example in this paper. The next logical extension of individual-based models in the near future is to simulate three or more species in complex environments.

While the individual-based approach is intuitively appealing, it can be difficult to put into practice, especially for full life cycle models. Information required for model formulation can exceed available data. Size data are readily available for many species. However, obtaining precise estimates of densities and vital rates, especially mortality, are problematic. Quantitative information on behavioral responses in nature (e.g. movement to avoid predation or to feed in a prey patch) are difficult to obtain. Once for-mulated, data required for rigorous model corroboration are often limited. Of course, the same arguments also apply, to varying degrees, to any mul-tispecies, and perhaps any single species, modeling effort. An individual-based model, like any model, is only as good as the information upon which it is based. Insufficient information for complete model formulation and testing is unavoidable. If we knew everything about a population or community, we would not need a model of its dynamics.

8.4 MULTISPECIES EXAMPLE

An individual-based model of yellow perch and walleye dynamics in Oneida Lake is used to illustrate:

(a) the individual-based approach applied to a multispecies situation;
(b) the importance of interspecific considerations for understanding popu-lation responses;
(c) the critical role computing power will probably play once again in multispecies modeling.

8.4.1 Percids and Oneida Lake

The population dynamics and trophic interactions of walleye and yellow perch in Oneida Lake, a large $(207\,km^2)$ eutrophic lake in central New York, have been studied for the past 40 years (Mills *et al.*, 1987; Mills and Forney, 1988). In most years, yellow perch were the dominant prey of walleye, while white perch, gizzard shad, and young walleye prey were of secondary importance. Observed yellow perch recruitments correlated with wind and temperature regimes during the larval stage, and with abundance, density-dependent growth, biomass of other YOY forage species, and size-selective predation by walleyes and adult perch during the juvenile and yearling stages. Observed walleye recruitments correlated with larval abundance, abundance and size of walleye forage (principally YOY and yearling yellow perch), and walleye cannibalism. Walleye are stocked annually as hatchings and support a large recreational fishery.

Major shifts in the abundance of available walleye forage that occurred during the period of record (1956–1988) were associated with changes in the population dynamics of yellow perch and walleye. The 'mayfly' period (1956–1969) was characterized by high densities of mayflies, and strong, stable year classes of walleye and yellow perch. The 'baseline' period (1970–1978) was characterized by absence of mayflies, relatively low abundances of white perch, and lower and less-stable recruitments of yellow perch and walleye. The 'forage fish' period (1979–1988), characterized by strong year classes of YOY white perch and gizzard shad, exhibited the lowest mean yellow perch and walleye recruitments of the three periods. For all three periods, similar adult abundances of yellow perch and walleye were observed, with year to year variability lower during the mayfly and forage fish periods than in the baseline period. In this paper, we use the model under baseline period conditions without walleye stocking. Rose *et al.* (in prep-a) compare model predictions for the baseline period to those for the forage fish and mayfly periods, as well to observed data for the three periods.

8.4.2 Model description

Overview

The model simulates the dynamics of yellow perch and walleye populations in Oneida Lake and is described in detail by Rose *et al.* (in prep-a). The model begins with spawning of individual females of each species and simulates growth and survival of each female's progeny as they develop through successive life stages (egg, yolk-sac larva, feeding larva, YOY

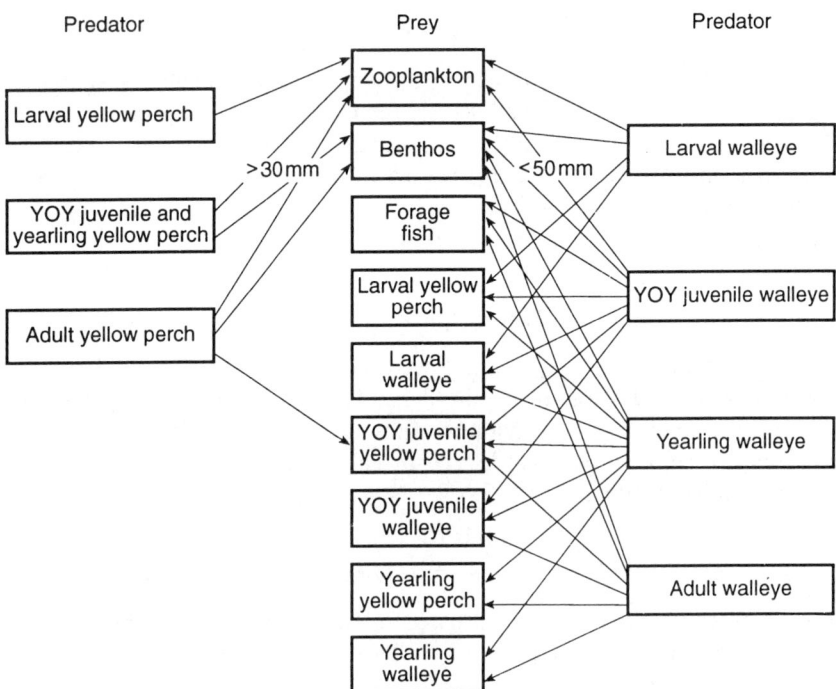

Fig. 8.1 Predator–prey interactions permitted in the individual-based yellow perch–walleye model as defined by nonzero vulnerability factors.

juvenile, yearling, and adults (age-2 and older)). Eggs and yolk-sac larvae are followed as cohorts from each female's spawn. The number and degree of development of eggs and yolk-sac larvae in each cohort are tracked daily. Individuals are followed for the larval, YOY juvenile, yearling, and adult life stages. Yellow perch are followed until age 10 and walleye until age 12. For individuals, the model keeps track of length (mm TL), weight (g wet weight), age, and life stage. Predator–prey relationships represented in the model are complex (Figure 8.1).

The model represents daily dynamics in a single, well-mixed compartment 260 m by 260 m by 6.8 m deep, scaled to represent Oneida Lake. The environment in the box consists of daily water temperature and the daily dynamics of a variety of prey types. Model simulations are for 200 years using a daily time step. Each year is 365 days long beginning on 10 April, just prior to the spawning seasons of both species.

Spawning

On each 10 April, the number and fecundities of individual spawning
female yellow perch and walleye are determined from the adult model.
Fraction mature and fecundity are specified as functions of female size.
Each female is assigned a temperature of spawning from species-specific
probability distributions; a female spawns on the first day that simulated
temperature exceeds the temperature of spawning.

Eggs and yolk-sac larvae

Each female's spawn of eggs is followed as a cohort. Day of hatching and
day of first feeding are temperature-dependent and determined by accumu-
lating daily fractional development rates until the cumulative exceeds 1.
Numbers of eggs or yolk-sac larvae in each cohort are reduced daily by
specified mortality rates.

Feeding life stages: initial conditions, development and age

All first-feeding larvae are assigned identical initial lengths: 6.6 mm
for yellow perch and 9 mm for walleye. Development from larvae to
YOY juveniles (metamorphosis) occurs at 20 mm for both species. Each
10 April, surviving individuals of each year class are promoted to the next
age.

Feeding life stages: growth and feeding

Daily growth in weight (W, g) is represented by the difference form of a
bioenergetics equation

$$W_t = W_{t-1} + p \times C_{\max} - E - R_{\text{tot}} \qquad (8.1)$$

where p is the proportion of $C_{\max}$ realized, $C_{\max}$ is maximum consumption
rate, E is egestion rate, excretion rate and SDA losses, and R_{tot} is total
metabolic rate. All rates are g g^{-1} day^{-1} in wet weight. We use parameter
values from previously published bioenergetics models (Rose *et al.*, in prep-
a). Maximum consumption $C_{\max}$ and metabolism R_{tot} depend on an individual's weight and water temperature. Egestion, excretion, and SDA are
lumped into the single loss term E, which is computed as a fraction of
realized consumption. An individual's weight is updated based on growth
and then its new length is determined using length–weight relationships.
Fish are allowed to lose weight but not to lose length.

Values of p are determined for each individual as realized consumption C_r

divided by $C_{\max}$. Realized consumption C_r is based on a type-II functional response relationship and is dependent on prey densities, vulnerability of each prey type to the individual predator, and half-saturation constants governing the rate of predator saturation. Daily consumption by each individual predator (C_r, g day^{-1}) and the consumption of each prey type j (C_j, g day^{-1}) is

$$C_r = \sum_{j=1}^{n} C_j \tag{8.2}$$

$$C_j = \frac{C_{\max} \times W \times \dfrac{(D^V_{ij})}{K_{ij}}}{1 + \sum_{k=1}^{n} \dfrac{(D^V_{ik})}{K_{ik}}} \tag{8.3}$$

where D_{ij}^V is the density of prey type j adjusted for its vulnerability to individual i, and K_{ij} is the half-saturation constant for individual i feeding on prey type j. Vulnerability-adjusted prey densities are obtained by either multiplying densities by a factor less than one (for zooplankton and benthos), or by determining the vulnerability of each individual fish prey item (forage fish, yellow perch, and walleye) based on its length relative to the predator length. Values of K are specified for size intervals of the predator and each prey type and were determined by calibration.

A total of seven prey types are represented that encompass the dominant prey eaten by yellow perch and walleye from first feeding through the adult stage. Zooplankton, benthos, and forage fish are each represented by one species: zooplankton by *Daphnia*, benthos by chironomids, and forage fish by YOY white perch. The remaining four prey types, YOY and yearling life stages of yellow perch and walleye, are simulated as individuals in the model. Prey densities are expressed as: number per liter for zooplankton, number per square meter for benthos, and number per cubic meter for fish.

Weights per individual of each prey type, and lengths for fish prey types, are either explicitly defined or simulated. Prey weight is used to convert between biomass and numbers consumed. Prey length is used when fish are prey to determine the vulnerability of an individual fish prey item to a particular predator. Each day the densities of zooplankton, benthos, and forage fish are updated based on numbers consumed and a specified replenishment rate; individuals of yellow perch and walleye eaten as prey are 'killed' and removed from the population.

The predator–prey interactions permitted in the model, as defined by non-zero vulnerability factors, are shown in Figure 8.1. Yellow perch can only eat YOY yellow perch whose length is less than 0.2 of the predator length. Only fish prey individuals whose length is less than 0.6 of the walleye predator length are vulnerable.

Feeding life stages: mortality

Larval mortality is specified at fixed rates for each species. A larva dies if a generated random number between 0 and 1 is less than its probability of dying. In addition, a few larvae die from encounters with YOY or yearling walleye. Mortalities of juvenile and yearling perch and walleye are computed from the consumption rates of walleye and perch. Predation by walleye and perch accounts for most of the mortality of YOY juvenile and yearling stages in Oneida Lake. Adult (age-2 and older) mortality of perch and walleye is density-dependent and represents fishing and natural mortality. Perch mortality increases with increasing numbers of perch adults. Walleye mortality decreases with increasing number of YOY perch, walleye, and forage fish because walleye are less apt to be caught when their natural forage is available.

8.4.3 Baseline simulation

Model predictions of yellow perch and walleye dynamics under walleye stocking were reasonable compared to observed values, and predictions without stocking deviated from observed as expected (Table 8.1; Figure 8.2; Rose *et al.*, in prep-a). Absence of walleye stocking resulted in higher densities of yellow perch (except for YOY Aug) and lower densities of walleye, which in turn, through density-dependent growth, caused smaller mean lengths at age of yellow perch and larger mean lengths at age for walleye. Perch and walleye adults (age-2 and older) in the absence of walleye stocking exhibited predator–prey cycling (Figure 8.3(a)). Rose *et al.* (in prep-a) perform a more critical comparison of the model predictions to Oneida Lake data.

8.4.4 Effects of changing adult mortality

To illustrate the importance of multispecies considerations, we separately varied adult perch and adult walleye mortality in the model under conditions of no walleye stocking. The monthly survival rate for each species was varied $\pm 4\%$ from baseline values, resulting in four simulations plus the baseline simulation. The following three model predictions were compared among the five simulations:

(a) number of adults (age-2 and older) in the model box at the beginning of each year (Figure 8.3);

Table 8.1 Predicted walleye and yellow perch mean densities (number per hectare) at various dates and ages for baseline simulations with and without walleye stocking, compared to observed values. Minimum and maximum observed values are shown in parentheses. Mean, minimum, and maximum observed values are computed based on Oneida Lake data for the baseline period (1970–1978). Mean model predictions are computed for years 20 through 200 of the simulation (see Rose *et al.* (in prep-a) for details)

| | Yellow perch | | | Walleye | | |
| | | Baseline simulation | | | Baseline simulation | |
Age/date	Observed	Stocking	No stocking	Observed	Stocking	No stocking
YOY Aug	11 900.0 (1200.0–37 700.0)	5909.2	4637.6	61.0 (2.0–144.0)	61.6	53.7
YOY Oct	2050.0 (100.0–8990.0)	2074.0	2468.4	40.0 (2.0–99.0)	46.9	39.9
Age-1 May	463.0 (5.0–3380.0)	1072.4	1683.7	43.0 (1.5–124.0)	37.6	34.0
Age-1 Sept	42.0 (1.0–224.0)	82.9	299.2	11.6 (5.0–400.0)	22.8	21.6
Age-2 Sept	11.7 (1.0–45.0)	43.9	137.8	–	–	–
Age-3 May	34.0 (0.9–144.4)	57.0	218.9	–	–	–
⩾Age-3 May	114.0 (80.0–200.0)	83.8	153.1	–	–	–
Age-4 May	–	–	–	7.1 (1.1–30.9)	21.2	20.1
⩾Age-4 May	–	–	–	29.2 (9.0–41.5)	25.6	24.4

(b) the mean and coefficient of variation (CV) of adult densities (Table 8.2);
(c) mean lengths at age (Figure 8.4).

The first 19 years of each simulation were ignored to eliminate initial condition effects. Mean and CV of densities and mean lengths at age were computed using 181 values (years 20 to 200).

The effects on adult densities of changing adult mortality of yellow perch and walleye were not symmetric. Increasing or decreasing adult perch mortality directly affected perch adult densities but had little effect on walleye adult densities. However, changing adult walleye mortality affected the adult densities of both walleye and perch (Figure 8.3; Table 8.2). Interannual variation in adult densities (CV) of walleye were unaffected by

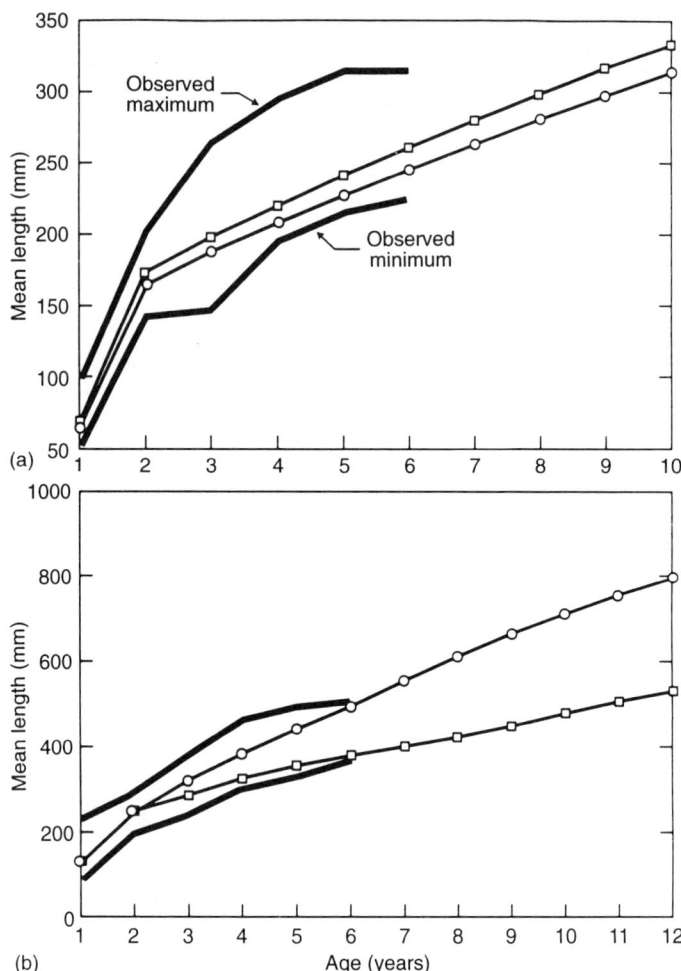

Fig. 8.2 Predicted mean lengths at age for (a) yellow perch and (b) walleye for baseline conditions (□) with and (○) without walleye stocking, compared to observed values. Observed values are shown as the minimum and maximum mean lengths at age for the baseline period.

changes in adult mortality, and were reduced for yellow perch under decreased mortalities of both species (Table 8.2).

Mean lengths at age for both species were also affected by changes in walleye mortality, while mean lengths at age of yellow perch were more affected than walleye lengths by changes in yellow perch adult mortality

(Figure 8.4). Density compensation was apparent for both species. Mean lengths at age of each species increased when their own adult mortality rate was increased and decreased when their adult mortality rate was decreased. However, as with adult densities, walleye lengths varied only slightly in response to changes in yellow perch mortality, while yellow perch lengths varied substantially in response to changes in walleye adult mortality. Because maturity and fecundity are size-dependent for both species, changes in mean length at age affected spawning.

Thus, the effects of changes in adult mortality on the population dynamics of yellow perch and walleye would not be obvious from single-species considerations. A simple view of the predator–prey relationship would imply that both species affect each other. Model analysis using a detailed individual-based model showed that changes in walleye mortality affected both yellow perch and walleye, while changes in yellow perch mortality affected yellow perch much more than walleye. The ability of both species to exhibit density compensation in growth (Figure 8.4), with its consequences on maturation and fecundity, coupled with their complex predator–prey interactions (Figure 8.1), caused asymmetric responses of each species to changes in the other species' mortality rates.

8.4.5 Computing considerations

A variety of numerical approximations were required for the individual-based yellow perch–walleye model to run with current desktop computing power. Following the millions of individual perch and walleye in the 20 700 ha Oneida Lake is not possible; we therefore simulated a 6.8 ha box. However, this resulted in the potential problems of large memory requirements to initially follow the thousands of first feeding larvae and of too few recruits surviving to adulthood in the box for some years. To deal with the problems of excessive memory requirements and too few surviving individuals, we developed a technique called resampling (Rose *et al.*, 1993b). A fraction of individuals are selected from spawners such that the relative proportions of first feeders coming from each female are maintained. Each model individual initially represents some number of identical population individuals. Resampling reduces the number of population individuals associated with each model individual permitting a fixed number of model individuals to be simulated while population numbers decline due to mortality. Thus, despite the emphasis on following individuals in model simulations, computing limitations forced the adoption of the resampling algorithm, with the consequence that all model individuals are not truly individuals but each represents some number of other identical population individuals.

Model individuals representing more than one population individual

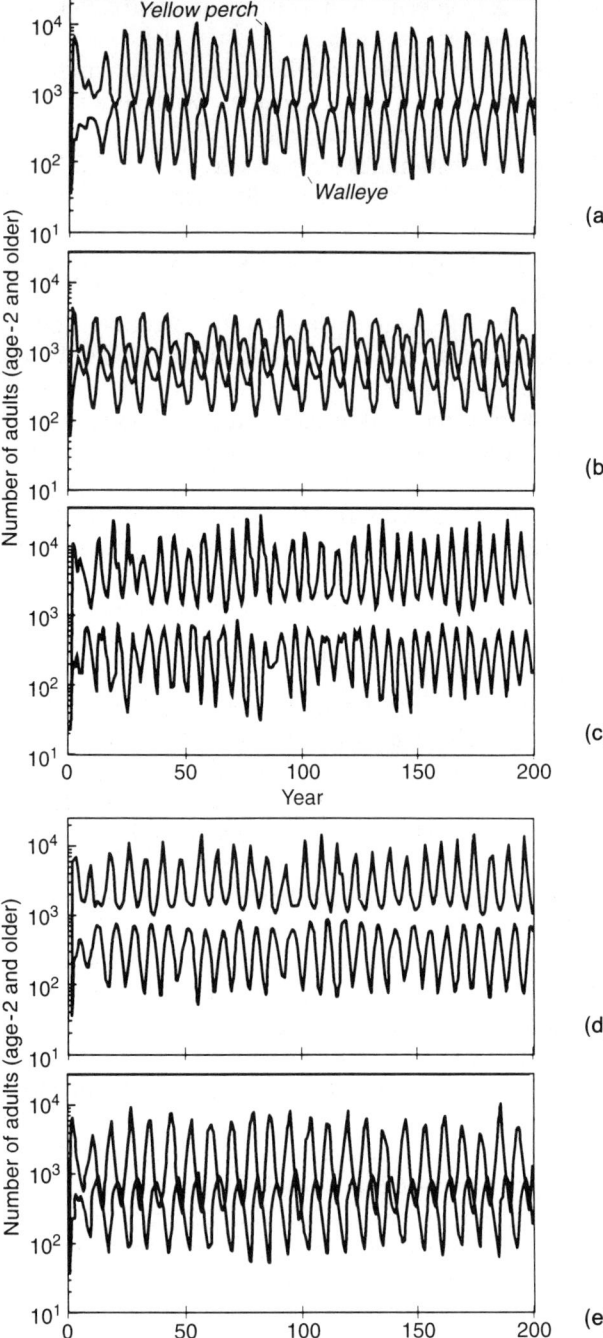

greatly complicated the simulation of predator–prey encounters and resulting mortality. Much of model development time (algorithm development and coding) was spent ensuring that predator–prey encounters and mortality rates based on model individuals were correctly scaled up to population individuals. Simulating predator–prey encounters when every individual is represented is straightforward. However, the computing limitations that forced us to use resampling resulted in complex calculations to relate encounters among model individuals to encounters among population individuals.

Even with the use of a reduced box size and a resampling algorithm, the yellow perch–walleye individual-based model is computing intensive. To illustrate, generation of the five 200-year simulations presented in this paper (baseline plus four with varied adult mortality) required about 100 h of CPU time on a DEC Alpha workstation and involved 474 million evaluations of the daily growth and death of model individuals. This number of 500 mm adult walleye lined up head-to-tail would circumspect the equator six times.

8.5 SUMMARY AND FUTURE DIRECTIONS

This chapter briefly reviewed the role of computing in multispecies modeling of fish populations. The availability of computing in the 1970s caused an initial surge in the development of multispecies models by fostering numerical (rather than analytical) solutions of complex systems of equations. Our review showed that, while computing power has increased exponentially since then, multispecies models used today have not changed much from those initial models. Computing considerations no longer appear to play a major role in affecting the formulation or running of classical multispecies models. The lack of apparent development in multispecies models is not because of a lack of interest. The issues requiring multispecies modeling have not diminished and, in fact, are better recognized now than ever. The importance of multispecies considerations for effective fishery management is critical given the generally low status of many exploited stocks. Further, many large-scale issues, such as biodiversity and

---

Fig. 8.3 Predicted number of adult (age-2 and older) yellow perch, upper traces, and walleye, lower traces, in the 6.8 ha model box on 10 April of each year for baseline conditions and simulations with adult monthly mortalities varied ±4%: (a) baseline, (b) decreased walleye mortality, (c) increased walleye mortality, (d) decreased yellow perch mortality, and (e) increased yellow perch mortality. All simulations were performed without walleye stocking.

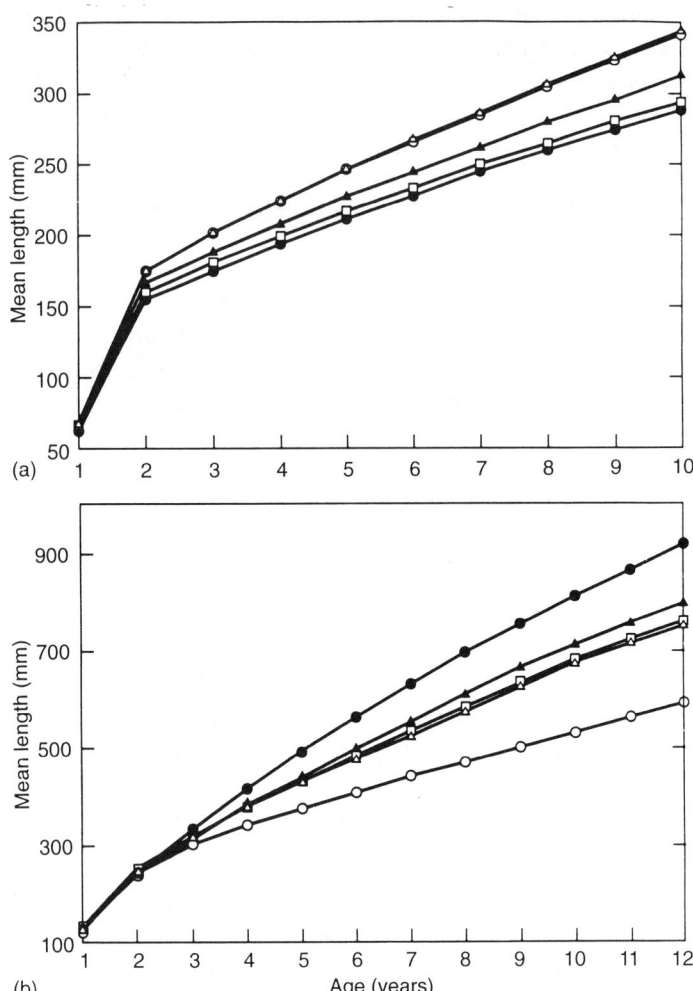

Fig. 8.4 Predicted mean lengths at age for adult (a) yellow perch and (b) walleye for baseline conditions and simulations with adult monthly mortality of each species varied ±4%: ▲ baseline, ○ decreased walleye mortality, ● increased walleye mortality, □ decreased yellow perch mortality, △ increased yellow perch mortality.

global change, are inherently multispecies in nature because they involve species replacement and composition changes. Multispecies models have not changed much from those initial applications because we still lack knowledge about most of the biological interactions underlying any multispecies modeling effort. Individual-based modeling offers a new alter-

Table 8.2 Predicted mean and CV of adult (age-2 and older) densities (number per hectare) for the Baseline simulation, and the simulations with walleye and yellow perch adult monthly mortalities varied $\pm 4\%$. All simulations were performed without walleye stocking.

	Mean		CV (%)	
Simulation	Yellow perch	Walleye	Yellow perch	Walleye
Baseline	374.2	66.2	98.1	62.0
Decreased walleye mortality	172.4	98.9	81.7	67.9
Increased walleye mortality	647.5	42.5	97.7	62.4
Decreased yellow perch mortality	520.1	60.5	88.5	62.3
Increased yellow perch mortality	283.7	64.3	101.9	62.4

native to multispecies modeling that may allow for further model developments.

We foresee a renewed role for computing in multispecies analyses with the increasing popularity, and apparent usefulness, of the individual-based approach to modeling. We used a yellow perch–walleye model of Oneida Lake to illustrate the individual-based approach applied to a multispecies situation. The importance of interspecific considerations for understanding population responses was demonstrated by the asymmetric responses of each species to changes in their adult mortality rates. The critical role of computing in these analyses was made apparent by the need for several approximations (small box, resampling) to allow execution of the model even on today's fast workstations. These approximations then required the addition of some complex computations to balance memory requirements with the need to obtain an adequate number of survivors and to ensure that predator–prey encounters were correctly scaled from model individuals to the population. The yellow perch–walleye model can be considered relatively simple in the sense that only two species are simulated in a spatially homogenous environment. As the individual-based approach is applied to more complex situations (e.g. three or more species, spatially explicit environments), the role of computing will increase even more.

We do not foresee major advances in software development. Most multispecies modeling efforts still use programming languages such as FORTRAN and BASIC. A few models use a higher-level language such as DYNAMO (Strange *et al.*, 1993) or use specialized software packages such as ECOPATH for budget modeling (Polovina, 1984), but these are the exception rather than the rule. There has been a lot of interest recently in object-oriented programming (Sequeira *et al.*, 1991), such as C++. While we see object-oriented programming as being helpful for code development, we do not see it as the panacea that will permit the unexperienced or uninitiated to develop complex models. Model coding will still remain the domain of the computer literate. Simple models for teaching and demonstration of ecological principles can be presented in easy-to-use software packages. However, our experience has been that models used for site-specific analyses or assessment always require a substantial amount of customization and continually evolve, making general software packages inefficient and unusable.

We do foresee hardware developments that could significantly impact on individual-based modeling of multispecies systems. Computing power based on sequential computations is still increasing exponentially. Furthermore, parallel architecture, in which multiple processors perform calculations simultaneously (Haefner, 1992), offers orders of magnitude increases in computing power. For the power of parallel computing to be realized, computations at each processor should be large compared to between-processor communication. The individual-based approach can be well-suited for parallel computing, as long as groups of model individuals can be identified that can be evaluated sequentially (for a time step or spatial region) without updating of their status or the environmental conditions they experience. Realizing the benefits of parallel computing is not necessarily trivial. To utilize presently available parallel machines requires explicit programming and modifications to model codes. Furthermore, some individual-based models are not obviously convertible to parallel computing. For example, in the yellow perch–walleye individual-based model presented here, updating of model individuals occurs after each predator is evaluated (rather than at the end of the timestep) to prevent prey from being eaten more than once and predators from eating more total prey than are available. While converting these calculations to utilize parallel architecture is possible, it does not simply involve parsing out groups of model individuals to different processors. In addition to major advances in parallel-based architectures, we also hope that software developments that will permit parallel machines to be used by computer-literature ecologists, not just the computer elite, will not lag too far behind.

In conclusion, computers were instrumental in the initial efforts at multispecies modeling of fisheries and, after playing little developmental role for almost two decades, will probably become important again with the increasing popularity of individual-based modeling. We attempted to demonstrate this with a brief review of past multispecies modeling efforts and an analysis of a yellow perch–walleye individual-based model. We believe that major computing advances will occur in hardware such as parallel architecture, rather than in software developments. Some of the ideas presented in this chapter are defensible, while others are opinions and conjecture. We hope the latter will stimulate further interest on the important topic of multispecies modeling of fish populations, especially on the utilization of the individual-based approach and the role of computers in such analyses.

ACKNOWLEDGEMENTS

This research was sponsored by the Electric Power Research Institute under contract RP2932-2 (DOE No. ERD-87-672) with the US Department of Energy, under contract DE-AC05-84OR21400 with Martin Marietta Energy Systems, Inc. This is Publication No. 4468 of the Environmental Sciences Division, Oak Ridge National Laboratory.

REFERENCES

Adams, S.M. and DeAngelis, D.L. (1987) Indirect effects of early bass-shad interactions on predator population structure and food web dynamics, in *Predation: Direct and Indirect Impacts on Aquatic Communities*, (eds W.C. Kerfoot and A. Sih), University Press of New England, Hanover, New Hampshire, pp. 103–17.

Allen, P.M. and McGlade, J.M. (1986) Dynamics of discovery and exploitation: the case of the Scotian Shelf groundfish fisheries. *Canadian Journal of Fisheries and Aquatic Sciences*, **43**, 1187–200.

Andersen, K.P. and Ursin, E. (1977) A multispecies extension to the Beverton and Holt theory of fishing, with accounts of phosphorus circulation and primary production. *Meddelelser fra Danmarks Fiskeri-og Havundersogelser*, **7**, 319–435.

Anon. (1990) Report of the Multispecies Assessment Working Group. *ICES C.M. 1991/Assess:7*.

Atkinson, M.J. and Griggs, R.W. (1984) Model of a coral reef ecosystem, II. Gross and net benthic primary production at French Frigate Shoals. *Coral Reefs*, **3**, 13–22.

Bax, N.J. (1985) Application of multi- and univariate techniques of sensitivity analysis to SKEBUB, a biomass-based fisheries ecosystem model, parameterized to Georges Bank. *Ecological Modelling*, **29**, 353–82.

Bax, N.J. (1991) A comparison of the fish biomass flow to fish, fisheries, and mammals in six marine ecosystems. *ICES Marine Science Symposia*, **193**, 217–24.

Bax, N.J. and Eliassen, J.-E. (1990) Multispecies analysis in Blasfjord, northern Norway: solution and sensitivity analysis of a simple ecosystem model. *Journal du Conseil International pour l'Exploration de la Mer*, **47**, 175–204.

Beyer, J.E. and Laurence, G.C. (1980) A stochastic model of larval fish growth. *Ecological Modelling*, **8**, 109–32.

Brander, K. (1988) Multispecies fisheries of the Irish Sea, in *Fish Population Dynamics*, 2nd edn, (ed. J.A. Gulland), John Wiley and Sons, New York, pp. 303–28.

Brander, K.M. and Mohn, R.K. (1991) Is the whole always less than the sum of the parts? *ICES Marine Science Symposia*, **193**, 117–19.

Brugge, W.J. and Holden, M.J. (1991) Multispecies management: a manager's point of view. *ICES Marine Science Symposia*, **193**, 353–8.

Clark, M.E. and Rose, K.A. (in press) Individual-based model of sympatric populations of stream resident rainbow trout and brook char: model description and baseline simulations.

Cohen, E.B., Grosslein, M.D., Sissenwine, M.P., Steimle, F. and Wright, W.R. (1982) Energy budget of Georges Bank, in Multispecies Approaches to Fisheries Management Advice (ed. M.C. Mercer), *Canadian Special Publication for Fisheries and Aquatic Sciences*, **59**, 95–107.

Crowder, L.B. (1988) Food web interactions in lakes, in *Complex Interactions in Lake Communities*, (ed. S.R. Carpenter), Springer-Verlag, New York, pp. 141–60.

Crowder, L.B., Rice, J.A., Miller, T.J. and Marschall, E.A. (1992) Empirical and theoretical approaches to size-based interactions and recruitment variability in fishes, in *Individual-based Models and Approaches in Ecology*, (eds D.L. DeAngelis and L.J. Gross), Chapman & Hall, New York, pp. 237–55.

Daan, N. and Sissenwine, M.P. (eds) (1989) Multispecies Models Relevant to Management of Living Resources. *ICES Marine Science Symposium*, **193**.

DeAngelis, D.L. and Gross, L.J. (eds) (1992) *Individual-based Models and Approaches in Ecology*, Chapman & Hall, New York.

DeAngelis, D.L., Godbout, L. and Shuter, B.J. (1991) An individual-based approach to predicting density-dependent dynamics in smallmouth bass populations. *Ecological Modelling*, **57**, 91–115.

DeAngelis, D.L., Rose, K.A. and Huston, M.A. (1994) Individual-oriented approaches to modeling populations and communities. *Lecture Notes in Biomathematics*, **100**, 390–410.

DeVries, D.R. and Stein, R.A. (1990) Manipulating shad to enhance sport fisheries in North America: an assessment. *North American Journal of Fisheries Management*, **10**, 209–23.

Fausch, K.D. (1988) Tests and competition between native and introduced salmonids in streams: what have we learned? *Canadian Journal of Fisheries and Aquatic Sciences*, **45**, 2238–46.

Fogarty, M.J., Sissenwine, M.P. and Grosslein, M.D. (1987) Fish population dynamics, in *Georges Bank*, (eds R.H. Backus and D.W. Bourne), MIT Press, Cambridge, MA, pp. 493–507.

Grigg, R.W., Polovina, J.J. and Atkinson, M.J. (1984) Model of a coral reef ecosystem, III. Resource limitation, community regulation, fisheries yield and resource management. *Coral Reefs*, **3**, 23–7.

Grosslein, M. *et al.* (1983) Ecosystem modeling as a fisheries management tool, in *Marine Ecosystem Modeling: Proceedings from a Workshop*, (ed. K.W. Turgeon), US National Oceanic and Atmospheric Administration, Washington, DC, pp. 235–42.

Gulland, J.A. (1989) Fish populations and their management. *Journal of Fish Biology (Suppl. A)*, **35**, 1–9.

Gulland, J.A. and Garcia, S. (1984) Observed patterns in multispecies fisheries, in *Exploitation of Marine Communities*, (ed. R.M. May), Springer-Verlag, Berlin, pp. 155–90.

Hackney, P.A. and Minns, C.K. (1974) A computer model of biomass dynamics and food competition with implications for its use in fishery management. *Transactions of the American Fisheries Society*, **103**, 215–25.

Haar, R., Swartzman, G. and Zaret, T. (1981) Evaluation of simulation models in power plant impact assessment: a case study using Lake Keowee, Report NUREG/CR-2436, US Nuclear Regulatory Commission, Washington, DC.

Haefner, J.W. (1992) Parallel computers and individual-based models: an overview, in *Individual-based Models and Approaches in Ecology*, (eds D.L. DeAngelis and L.J. Gross), Chapman & Hall, New York, pp. 126–64.

Hennemuth, R.C. (1979) Man as predator, in *Contemporary Quantitative Ecology and Related Econometrics*, (eds G.P. Patil and M. Rosenzweig), International Co-operative Publishing House, Fairland, Maryland, pp. 507–32.

Hinckley, S., Hermann, A.J. and Megrey, B.A. (in press) Development of a spatially explicit, individual-based model of marine fish early life history, II. Methods. *Marine Ecology Progress Series*.

Huston, M.A., DeAngelis, D.L. and Post, W.M. (1988) New computer models unify ecological theory. *BioScience*, **38**, 682–91.

Jager, H.I. *et al.* (1993) An individual-based model for smallmouth bass reproduction and young-of-year dynamics in streams. *Rivers*, **4**, 91–113.

Jarre, A., Muck, P. and Pauly, D. (1991) Two approaches for modelling fish stock interactions in the Peruvian upwelling ecosystem. *ICES Marine Science Symposia*, **193**, 171–84.

Kerr, S.R. and Ryder, R.A. (1989) Current approaches to multispecies analyses of marine fisheries. *Canadian Journal of Fisheries and Aquatic Sciences*, **46**, 528–4.

Kitchell, J.F. (ed.) (1992) *Food Web Management: A Case Study of Lake Mendota*, Springer-Verlag, New York.

Kitchell, J.F. and Crowder, L.B. (1986) Predator-prey interactions in Lake Michigan: model predictions and recent dynamics. *Environmental Biology of Fishes*, **16**, 205–11.

Laevastu, T., Favorite, F. and Larkins, H.A. (1982) Resource assessment and evaluation of the dynamics of the fisheries resources in the northeastern Pacific with numerical ecosystem models, in *Multispecies Approaches to Fisheries Management Advice*, (ed. M.C. Mercer), *Canadian Special Publication for Fisheries and Aquatic Sciences*, **59**, 70–81.

Laevastu, T. and Larkins, H.A. (1981) *Marine Fisheries Ecosystem: Its Quantitative Evaluation and Management*, Fishing News Books Ltd, Farnham, England.

Laurec, A., Biseau, A. and Charuau, A. (1991) Modelling technical interactions. *ICES Marine Science Symposia*, **193**, 225–36.

Madenjian, C.P. and Carpenter, S.R. (1993) Simulation of the effects of time and size at stocking on PCB accumulation in lake trout. *Transactions of the American Fisheries Society*, **122**, 492–9.

May, R.M., Beddington, J.R., Clark, C.W., Holt, S.J. and Laws, R.M. (1979) Management of multispecies fisheries. *Science*, **205**, 267–75.

Mercer, M.C. (ed.) (1982) Multispecies Approaches to Fisheries Management Advice. *Canadian Special Publication for Fisheries and Aquatic Sciences*, **59**.

Mills, E.L. and Forney, J.L. (1988) Trophic dynamics and development of freshwater pelagic food webs, in *Complex Interactions in Lake Communities*, (ed. S.R. Carpenter), Springer-Verlag, New York, pp. 11–30.

Mills, E.L., Forney, J.L. and Wagner, K.J. (1987) Fish predation and its cascading effect on the Oneida Lake food chain, in *Predation: Direct and Indirect Impacts on Aquatic Communities*, (eds. W.C. Kerfoot and A. Sih), University Press of New England, Hanover, New Hampshire, pp. 118–31.

Murawski, S.A. (1984) Mixed-species yield-per-recruitment analyses accounting for technological interactions. *Canadian Journal of Fisheries and Aquatic Sciences*, **41**, 897–916.

Murawski, S.A. (1991) Can we manage our multispecies fisheries? *Fisheries*, **16**, 5–13.

Murawski, S.A., Lange, A.M. and Iodine, J.S. (1991) An analysis of technological interactions among Gulf of Maine mixed-species fisheries. *ICES Marine Science Symposia*, **193**, 237–52.

Pace, M.L., Glasser, J.E. and Pomeroy, L.R. (1984) A simulation analysis of continental shelf food webs. *Marine Biology*, **82**, 47–63.

Patten, B.C. (1969) Ecological systems analysis and fisheries science. *Transactions of the American Fisheries Society*, **98**, 570–81.

Patten, B.C. *(et al.)* (1975) Total ecosystem model for a cove in Lake Texoma, in *Systems Analysis and Simulation in Ecology* (ed. B.C. Patten), Academic Press, New York, Vol. III, pp. 205–421.

Patten, B.C. *et al.* (1983) Ecosystem modeling as an environmental management tool, in *Marine Ecosystem Modeling: Proceedings from a Workshop*, (ed. K.W. Turgeon), US National Oceanic and Atmospheric Administration, Washington, DC, pp. 243–55.

Paulik, G.J. (1969) Computer simulation models for fisheries research, management, and teaching. *Transactions of the American Fisheries Society*, **98**, 551–9.

Persson, L. and Diehl, S. (1990) Mechanistic individual-based approaches in population/community ecology of fish. *Annales Zoologici Fennici*, **27**, 165–82.

Ploskey, G.P. and Jenkins, R.M. (1982) Biomass model of reservoir fish and fish-food interactions, with implications for management. *North American Journal of Fisheries Management*, **2**, 105–21.

Polovina, J.J. (1984) Model of a coral reef ecosystem, I. The ECOPATH model and its application to French Frogate Shoals. *Coral Reefs*, **3**, 1–11.

Pope, J.G. (1989) Multispecies extensions to age-structured assessment models. *American Fisheries Society Symposium*, **6**, 102–11.

Pope, J.G. (1976) The effect of biological interaction on the theory of mixed fisheries. *International Commission for the Northwest Atlantic Fisheries Selected Papers*, **1**, 157–62.

Regier, H.A. and Hartman, W.L. (1973) Lake Erie's fish community: 150 years of cultural stresses. *Science*, **180**, 1248–55.

Rice, J.A., Miller, T.J., Rose, K.A., Crowder, L.B., Marshcall, E.A., Trebitz, A.S. and DeAngelis, D.L. (1993) Growth rate variation and larval survival: inferences from an individual-based size-dependent predation model. *Canadian Journal of Fisheries and Aquatic Sciences*, **50**, 133–42.

Riffenburgh, R.H. (1969) A stochastic model of interpopulation dynamics in marine ecology. *Journal of the Fisheries Research Board of Canada*, **26**, 2843–80.

Rose, K.A., Rutherford, E.S., SinghDermott, D., Forney, J.L. and Mills, E.L. (in prep-a) An individual-based model of walleye and yellow perch population dynamics in Oneida Lake: model description and effects of alternate walleye prey.

Rose, K.A. and Cowan, J.H. (1993) Individual-based model of young-of-the-year striped bass population dynamics: I. Model description and baseline simulations. *Transactions of the American Fisheries Society*, **122**, 415–38.

Rose, K.A., Cowan, J.H., Clark, M.E. and Houde, E.D. (in prep-b) Simulating bay anchovy population dynamics using an individual-based approach.

Rose, K.A., Tyler, J.A., Chambers, R.C., MacPhee, G. and Danila, D.J. (in press). Simulating winter flounder population dynamics using coupled individual-based young of the year and age-structured adult models.

Rose, K.A., Cowan, J.H., Houde, E.D. and Coutant, C.C. (1993a) Individual-based modelling of environmental quality effects on early life stages of fishes: a case study using striped bass. *American Fisheries Society Symposium*, **14**, 125–45.

Rose, K.A., Christensen, S.W. and DeAngelis, D.L. (1993b) Individual-based modeling of populations with high mortality: a new method of following a fixed number of model individuals. *Ecological Modelling*, **68**, 272–92.

Sharov, A.F. and Kriksunov, E.A. (1991) A simulation model of two interacting exploited fish populations through competition in the early phases of life. *ICES Marine Science Symposia*, **193**, 113–16.

Sequeira, R.A., Sharpe, P.J.H., Stone, N.D., El Zik, K.M. and Makela, M.E. (1991) Object oriented simulation: plant growth and discrete organ to organ interactions. *Ecological Modelling*, **58**, 55–89.

Silliman, R.P. (1969) Analog computer simulation and catch forecasting in commercially fished populations. *Transactions of the American Fisheries Society*, **98**, 560–9.

Sissenwine, M.P., Cohen, E.B. and Grosslein, M.D. (1984) Structure of the Georges Bank ecosystem. *Rapports et Proces-Verbaux des Reunions. ICES*, **183**, 243–54.

Sissenwine, M.P. (1984) Perturbation of a predator-controlled continental shelf ecosystem, in *Variability and Management of Large Marine Ecosystems*, (eds K. Sherman and L.M. Alexander), AAAS Selected Symposium 99, Westview Press, Boulder, Colorado, pp. 55–85.

Sparre, P. (1991) Introduction to multispecies virtual population analysis. *ICES Marine Science Symposia*, **193**, 12–21.

Strange, E.M., Moyle, P.B. and Foin, T.C. (1993) Interactions between stochastic and deterministic processes in stream fish community assembly. *Environmental Biology of Fishes*, **36**, 1–15.

Strobele, W.J. and Wacker, H. (1991) The concept of sustainable yield in multi-species fisheries. *Ecological Modelling*, **53**, 61–74.

Sutcliffe, W.H., Drinkwater, K. and Muir, B.S. (1977) Correlations of fish catch and environmental factors in the Gulf of Maine. *Journal of the Fisheries Research Board of Canada*, **34**, 19–30.

Tetra Tech (1980) Methodology for evaluation of multiple power plant cooling system effects. Vol. V: Methodology application to prototype-Cayuga Lake, Report EPRI EA-1111, Electric Power Research Institute, Palo Alto, CA.

Turgeon, K.W. (ed.) (1982) *Marine Ecosystem Modeling: Proceedings From a Workshop*, US National Oceanic and Atmospheric Administration, Washington, DC.

Tyler, J.A. and Rose, K.A. (1994) Individual variability and spatial heterogeneity in fish population models. *Reviews in Fish Biology and Fisheries*, **4**, 91–123.

Ulanowicz, R.E. (1986) *Growth and Development: Ecosystem Phenomenology*, Springer-Verlag, New York.

Van Winkle, W., Rose, K.A. and Chambers, R.C. (1993) Individual-based approach to fish population dynamics: an overview. *Transactions of the American Fisheries Society*, **122**, 397–403.

Volterra, V. (1928) Variations and fluctuations of the number of individuals in animal species living together. *Journal du Conseil International pour l'Exploration de la Mer*, **3**, 1–51.

Walsh, J.J. (1981) A carbon budget for overfishing off Peru. *Nature*, **290**, 300–4.

Walters, C.J., Hannah, C.G. and Thompson, K. (1992) A microcomputer program for simulating effects of physical transport process on fish larvae. *Fisheries Oceanography*, **1**, 11–19.

Wilson, J.A., French, J., Kleban, P., McKay, S.R. and Townsend, R. (1991a) Chaotic dynamics in a multiple species fishery: a model of community predation. *Ecological Modelling*, **58**, 303–22.

Wilson, J.A., Kleban, P., McKay, S.R. and Townsend, R.E. (1991b) Management of multispecies fisheries with chaotic population dynamics. *ICES Marine Sciences Symposia*, **193**, 287–300.

Yodiz, P. (1994) Predator–prey theory and management of multispecies fisheries. *Ecological Applications*, **4**, 51–8.

Chapter nine

Computers and the future of fisheries

Carl J. Walters

9.1 INTRODUCTION

The world's fisheries are a shambles. Everywhere we see signs of massive overexploitation, breakdown of regulatory and enforcement systems, and woefully inadequate investment in assessment and science. The credibility of fisheries science has been questioned with studies that have revealed severe overestimates in abundance and productivity in historical assessments for several major fisheries (Hutchings and Myers, 1994; McGuire, 1991; Parma, 1993; Pauly, 1994; Walters, 1994a), indicating that we have contributed directly to the overcapitalization that we have traditionally blamed on greed and managerial stupidity. And all this has happened while whole new worlds of information gathering and analysis have been opened to us through computer technology. I am reminded of an old poster that is still displayed prominently on my office wall, a gift from participants in a 1979 Sea Lamprey International Symposium, displaying the adage 'To err is human. To really foul things up requires a computer!' In those days the adage was referring to the growing pains of an information management industry for such human affairs as banking; little did we know at the time how well it would apply to models that were appearing at the time for improving fisheries assessments.

Everyone knows that fisheries are complex systems, involving interactions among fish, people, and aquatic ecosystems. But we still seem determined to pretend that fisheries biology is somehow at the heart of this complexity, and that our skills as biologists are the most important ones. Most fisheries students learn very quickly how wrong this pretense is when they take their first real job in fisheries, and are plunged into the world of trying to make useful assessments and quantitative predictions with messy

Computers in Fisheries Research.
Edited by Bernard A. Megrey and Erlend Moksness.
Published in 1996 by Chapman & Hall, London. ISBN 0 412 59550 8

data; most end up spending far more time handling spreadsheets than they do handling fish, and wishing that they had better understood the gibberish in their statistics and computer courses. What they are learning at this time is something very fundamental: many fisheries processes and problems are essentially statistical rather than biological. After all, what is a 'population process' other than a statistical summation of the effects of things that happen to or are exhibited by many individual animals? How could we hope to understand how such summations behave in space and time just by knowing much about the typical animal that engages in them, especially when we are also taught early on that individuals vary enormously? Computers are opening up whole new ways for us to monitor and represent the complex statistical dynamics of fishing systems, and we must make every effort to capitalize on this opportunity. Hopefully books like this one will hasten the difficult educational process that will make that possible.

This chapter reviews just a few of the opportunities and pitfalls that computers have brought to fisheries science. Emphasis in the following sections is on those computer applications most directly and immediately useful in managing fisheries better, rather than on applications that will give us more fundamental understanding in the long run. I believe in fact that working harder to answer direct and well-focused applied questions is the quickest way for us to come to grips with the fundamental questions as well.

9.2 OPENING NEW WINDOWS FOR MEASURING FISHERY DYNAMICS

Here are a few of the things that we can do today with computers, that would have been unthinkable or outrageously costly even a few decades ago (chapter numbers refer to topics discussed in this volume):

(a) Equip every vessel in a fishing fleet with transponders that accurately report positions every few minutes or hours to a central computer data system that can plot fishing distributions and pinpoint movements that may represent illegal fishing activities.

(b) Have fishermen report their catches and oceanographic observations, accurately georeferenced, through the same transponder systems so that every vessel in a sense becomes part of a survey or 'test fishing' operation (Chapter 5).

(c) Access massive oceanographic, meteorological, and biophysical ('habitat') databases for comparison with direct information from fishing vessels and biological surveys (Chapters 3, 4 and 6).

(d) Tag large numbers of fish with archival computer tags (or transmitting tags linked to our personal computers) that let us see how fish use the world, in remarkable detail.

(e) Compare complex atomic ratio, scale pattern, or DNA segment 'signatures' for large numbers of individual fish to reference signatures for particular stocks/areas of fish origin, to assess stock composition from fisheries where many stocks are taken together (Chapter 6).

(f) Use image processing methods to turn complex sonar, laser, and other physical signals into abundance and distribution maps (Chapter 6).

(g) Keep careful and accurate records of past management activities and measured responses of fish and fishermen to these activities.

(h) Run many population simulations with different parameter values, to determine the set of parameter combinations that are consistent with historical data (i.e. how large the uncertainty is about the parameter values).

(i) Simulate dispersal and migration patterns of thousands of small bins or packets of fish, to represent the spatial dynamics of fish relative to where and when fishing takes place (Chapters 7 and 8).

(j) Use information on ecological interactions (predation, competition, etc.) to construct multispecies and ecosystem models that can help in design of balanced policies for ecosystem management (i.e. stop treating fish stocks as independent of one another in harvest management planning) (Chapter 8).

(k) Present spatial and ecosystem simulations to fishery managers in the format of easily used simulation games, where they can do things like move fishing areas and times around on their PC screen map of the fishery, with a mouse (Chapters 3 and 7).

(l) Devise complex empirical rules for making forecasts and carrying out regulatory actions to achieve harvest goals (Chapter 2).

Consider the opportunities for assessment that these things have created. At the heart of all of them is the notion that we can now gather enough information to begin to discard the old and very misleading mental model of a fish stock as a dynamic pool, distributed like molecules in a chemical reactor vat and behaving with comparable simplicity. This model has got us into trouble over and over again in fisheries assessment. While lulling us into a sense of confidence that we are studying the right processes (growth, recruitment, mortality), it has left us largely unable to use information effectively and to prescribe wise policies at the scales in space and time where policy must actually be implemented. On the information side, we are encouraged by the pool model to view fishing as somehow distributed simply in space, and to use indices such as aggregate catch per effort to describe changes in relative abundance. On the prescriptive side, we are able to provide little more in the way of policy advice than crude estimates of sustainable yield, population growth/decline rates, and likely responses to management measures for varying the sizes of fish captured. What is fundamentally missing in the pool model is something that much concerns every biologist and fisherman: spatial structure and dynamics. Fish are not randomly distributed, and their ontogeny almost always involves a complex trajectory of dispersal and migration through a range of habitats and ecological interactions. It drives us to distraction that fishermen know such distributions and trajectories very well, and exploit them in complex ways to end up providing us with catch and catch rate data that tell us very little about what populations are doing when interpreted in crude (nonspatial) ways. Sometimes catch rates decline very rapidly as harvesting develops, as natural aggregations of fish are removed, wthout there really being much impact of fishing on the stock. In other cases, aggregations form fast enough in the face of harvest removals, and the range used by fish contracts after such removals, to make us think from crude statistics that everything is well when in fact the stock is rapidly collapsing. In still other cases, fisheries progressively mine out local populations further and further from ports, without this sequential depletion being evident at all in traditional statistics.

There are two information situations that should frighten every fisheries scientist. The first is where collapse is so obvious in the data that any fool can see it, and the frightening questions in this case are about how to reverse the trend and deal with socioeconomic hardship. The second is where things look really good (catches stable or growing, catch per effort stable), and fishery managers are smiling complacently about what a good job they are doing. Many are the times that I have sat in management planning meetings with such smiling people, while wanting to scream in frustration about the lack of even crude spatial statistics to help determine whether aggregations are shrinking or fishing grounds are being sequentially depleted. The reader might have noticed that these two information situations cover most of the world's fisheries; the few situations where we perhaps need not be frightened are ones where complex spatial structure (many lakes, streams, reefs, etc.) has offered many opportunities to test alternative management approaches without putting the whole system at risk.

Many fisheries today are moving to quota management systems so as to improve economic performance, simplify the manager's job, and presumably foster attitudes toward better resource husbandry by fishermen. However, these systems radically increase the demand placed on assessment science to provide more precise stock size estimates so as to avoid depensatory effects of quota removals (Pearse and Walters, 1992). In most fisheries we cannot even begin to meet this demand with the precision that managers and fishermen seem to expect, and I personally am coming to think of quota management as potentially one of the most destructive innovations in the history of fisheries development.

The dangers of quota management are not lost on fishermen and the public, and we are beginning to see more frequent calls to design regulatory policies that will absolutely close the door on overfishing by complementing quota management with space/time harvesting restrictions that will place a firm upper bound on fishing mortality rates (i.e. directly limit the proportion of fish that are exposed to harvest). The design of such refuge systems requires fairly detailed information about how fish are distributed and how they move about, and correspondingly detailed models for comparing alternative refuge sizes and timings. We have very little experience in fisheries with the development of such **regulatory tactics models**, except in the very special case of gauntlet fisheries for migrating Pacific salmon. I predict that development of detailed regulatory models, and the data needed to make them work, is going to be one of the most exciting and important areas of fisheries research over the next decade or two.

A cursory look at recent literature on fisheries modeling and assessment methods might give the impression that we are getting better at both modeling how fish populations behave and at linking data to models through elaborate estimation procedures that link diverse sources of data. Terms such as stock synthesis (Methot, 1990) and auxiliary information (Deriso *et al.*, 1985; Fournier and Warburton, 1989; this volume, Chapter 7) have come to connote the promise of assessment methodology that can keep up with the demands of quota management. To some degree this promise is being fulfilled, but the situation is not nearly as bright as it may appear. Many of the recent advances in modeling and data analysis are not in fact aimed at or capable of producing more precise assessments at all; what they are doing instead is providing more accurate and honest assessments of how grossly uncertain we really are about abundance and production. We are coming to question many assumptions of traditional models, ranging from whether production dynamics are stationary to whether some types of data carry any information at all. Techniques such as Bayesian estimation are allowing us to express the uncertainty represented by such questions in terms of probability distributions, and there has been a nasty tendency recently for such distributions to get broader rather than narrower. Such admissions of deep and persistent uncertainty will almost certainly help to fuel the demand noted in the previous paragraph for regulatory models and systems that directly reduce the risk of overharvest.

9.3 DESIGNING ROBUST POLICIES FOR LIVING WITH UNCERTAINTY

Although computer data acquisition and modeling systems may provide better stock assessments and understanding of fish–environment relations, it is likely that fisheries are always going to be managed under considerable uncertainty. This is particularly so in relation to impacts of environmental and climate change; even if we come to understand how such factors influence fish, there is little prospect for predicting accurately how the factors themselves will change. Over the past 20 years, computer simulation and optimization methods have allowed much progress in the design of feedback and adaptive policies for coping with unpredictable change; this policy design work has relied upon or assumed that we can describe and predict statistical patterns of variability even if we cannot say what causes that variability. Indeed, policy designs based on such statistical descriptions are robust in the sense that they are entirely invariant to causation of the variation (we get the same policy design answers no matter how we comfort ourselves through attempts to explain how the variation arises).

A particularly interesting recent finding has been about how classic, fixed exploitation rate (or fixed fishing rate F) policies may provide a robust way to deal with the effects of climate change on biological carrying capacities (Walters and Parma, 1995; Parma and Deriso, 1990). We already know that such policies are near optimum for dealing with risk-averse harvest management objectives (Deriso, 1985; Hilborn and Walters, 1992), but they may have another really important bonus as well even in maximizing long-term harvest objectives that do not involve risk aversion. Basically, it appears that successful implementation of a fixed harvest rate strategy should allow quite close tracking of changes in optimum population size associated with changes in carrying capacity. Figure 9.1 shows a 100-year simulation of a population that is subject to wide swings in carrying capacity (as measured by equilibrium unfished stock size). Spawning stock size under an optimum fixed exploitation rate policy (calculated from the slope at low stock sizes of the population's mean stock–recruit relation) moves up and down with the carrying capacity. But the really interesting thing is how this spawning stock size tracks the theoretical optimum spawning stock size, where this theoretical optimum is calculated by dynamic programming for each simulated year assuming perfect knowledge as of that year of all future carrying capacity changes. The tracking is not perfect, since the optimum policy anticipates big carrying capacity changes and begins to adjust stock size ahead of time if necessary to capitalize on such changes when they arrive. However, at least in cases such as Figure 9.1, where the environmental change is not too rapid, the fixed exploitation policy gives total long-term harvests that are 90% or higher of the theoretical optimum. This finding is not at all sensitive to the population model used in the calculation, except that the impact of the environmental change must be mainly on parameters that determine maximum population size (carrying capacity) rather than on the population's intrinsic rate of increase (slope of recruitment curve) when abun-

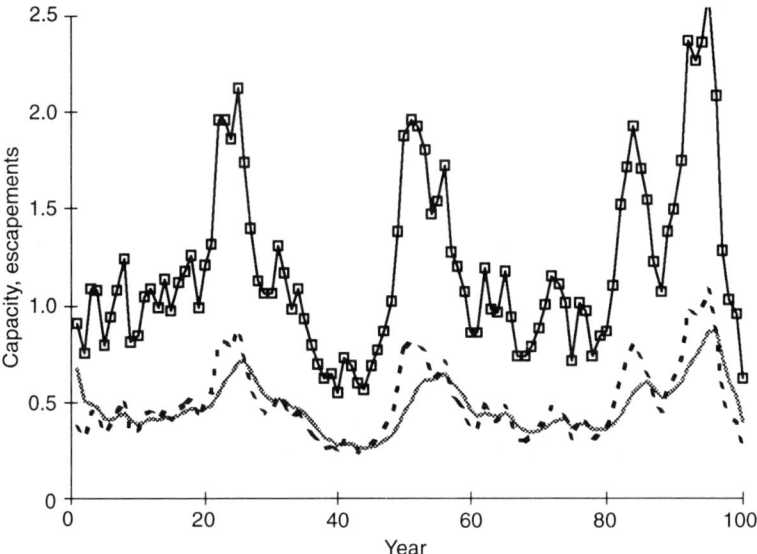

Fig. 9.1 Simulation showing that a fixed harvest rate strategy can result in population sizes that closely track the theoretical optimum for responding to changes in carrying capacity caused by climate change. The carrying capacity line (——) represents simulated change in unfished stock size due to climate effects (generated by simple autoregressive model). The fixed U line (····) represents stock size that would be achieved by taking the best constant fraction (U) of the stock every year. The optimum escapement line (– – –) represents stock size target that a manager with perfect knowledge of future climate change would try to achieve each year (computed using dynamic programming).

dance is low. When intrinsic rate of increase is also variable with environmental conditions, the harvest rate should be varied over time to track such changes.

Such simulations do not, of course, tell how to achieve a desired fixed harvest rate in terms of annual harvest regulation. In quota management systems, this will presumably involve setting the quota for each year to the target harvest rate times an estimate of stock size. In cases such as Pacific salmon, we can directly regulate risk of harvest through area/time closures to fishing. Much work needs to be done using simulations and careful statistical analyses of stock size estimation to determine whether target harvest rates under quota management should be adjusted downward to reflect uncertainty in stock size estimates. Preliminary indications from simulations of how quota setting errors may affect long-term management performance are that target harvest rates should be adjusted downward by

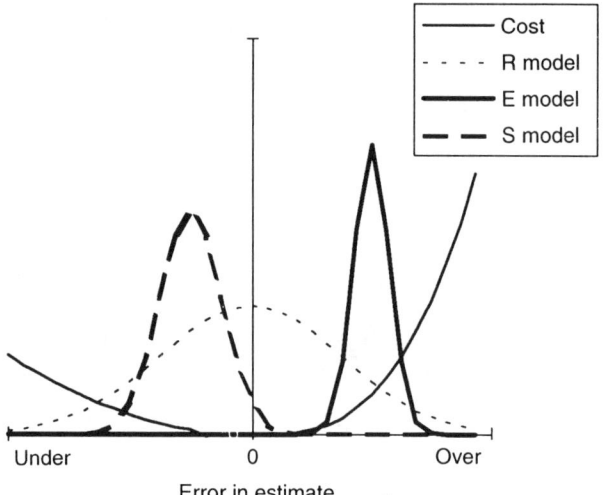

Fig. 9.2 Statistical variation in estimates of policy parameters such as maximum sustainable yield depends on the model used for estimation, and there is a cost of making estimation errors. Estimates can be quite variable (R case) when realistic models are used with uninformative data, resulting in high expected cost (high odds of estimates being far enough off to cause high cost). Estimates from very simple equilibrium models (E case) may be quite precise for the same data, but dangerously biased. An ideal population model for assessment (S case) often is a deliberately oversimplified model that is not too badly biased (and in a safe direction), but gives estimates that are less variable than would be achieved with a more realistic model.

a factor of roughly $\exp(CV^2)$, where CV is the coefficient of variation of the stock size estimate (Walters and Pearse, 1994).

9.4 KEEPING THE BAD GUYS AT BAY: SEEING HOW FISHERIES REALLY OPERATE

Illegal fishing activities are rapidly becoming one of the biggest threats to sustainability of the world's fisheries. Historically, fisheries scientists have not worried much about poaching, and it was not a big problem when fishermen had free run of the seas, lakes, and streams. But factors ranging from overfishing to development of Asian markets are now making fish more valuable, and our very efforts to create sustainable fisheries through quota management and harvest regulation are creating incentives for development of 'underground' fishing and marketing systems. At the same time, there is political pressure to cut back on public subsidies to fisheries, including such management costs as enforcing regulations, and to place this burden on fishing industries that in many cases cannot afford traditional, manpower-intensive enforcement systems. Field enforcement and management staff in almost every public agency now have horror stories to tell about well-organized illegal fishing activities, complex schemes for transporting fish to safe marketing locations, and routine violations of regulations and quota shares even by legally licensed fishermen.

It does not take much misreporting of catches to really foul up our existing stock assessment methods and models. For instance, we very nearly destroyed British Columbia's largest salmon stock in 1994, the Adams River dominant sockeye (*Oncorhynchus nerka*), because of an overestimate of its abundance at one point along the gauntlet of fisheries that it faces during its spawning migration. Very high catches were reported in one fishing area (the Johnstone Strait), and this led biologists to revise run size estimates upwards and to open the next fishing area (Fraser River mouth); luckily statistics from this next area showed that the stock had in fact already been overfished, and the area was closed in time to prevent an ecological disaster. The computer models for run reconstruction and estimation had apparently failed badly for the 1994 situation. Partly this was due to an unusual migration routing by the fish, but there are also suspicions that the very high catches reported in the Johnstone Strait did not in fact actually happen in the legal fishery, and instead represent fish caught illegally in various closed areas and times along the migration. In short, what looked to biologists like high abundance may actually have been a symptom of large-scale illegal fishing.

Computers can help in two dramatic ways to substantially reduce illegal fishing. The most obvious way is to require all fishing vessels to carry location devices, for which computers are necessary for interpreting transponder signals and providing maps of vessel distributions. But another thing we can do is to feed the great mass of information gathered this way into expert system software, programmed to look for and warn us about patterns of vessel movement that likely represent either illegal fishing or transport of fish. Fishermen are understandably not very happy about this concept (Big Brother is watching), but they may eventually come to support it if fisheries agencies can convince them of three key points:

(a) the seas and the fish are still considered public property, and the public has every right to monitor how these properties are used;
(b) their legal livelihoods really are being threatened by the bad guys;
(c) today's fishing technologies and transportation systems are so quick and efficient today that traditional enforcement methods cannot offer a credible threat of catching them in the act.

9.5 EXPLORING OPTIONS AND OPPORTUNITIES: INTEGRATING ASSESSMENT AND MANAGEMENT

In recent years there has developed a dangerous gap between the quantitative, computer-intensive world of fisheries assessment and the world of the practising fishery managers who work directly with industry and political decision makers. Many fishery managers understand neither the technology or limitations of the analysis results that they convey in various formats to fishermen and politicians, and there is often little onus on them to even keep up with recent literature pointing out alternative management approaches and difficulties. To some extent the misunderstandings that this state of affairs has created are beginning to be circumvented through development of direct, cooperative working arrangements between fishing industry representatives and assessment scientists.

It appears that one of the most effective ways to promote better communication and mutual education among these management players is to literally place them in game-playing situations (Walters, 1994b) where computer models are used to represent at least some of the possible impacts of alternative management strategies. Such games can be extremely useful tools for identifying imaginative new policy options and for understanding basic dynamic principles and limits, even if the models used in them are quite unrealistic. However, today we do not have to be content with crude models; spatial simulations and graphical visualization techniques can be used to make fisheries dynamics literally come alive on computer screens. Ideas that we used to try to explain with complex graphs, tables, or equations can now be conveyed in pictures that not only inform, but also help to stimulate new ideas.

I was recently treated to a taste of the potential power of management gaming entirely by accident, during a workshop with rock lobster fishermen and biologists in South Australia. We were trying to develop a messy spatial model for the stock (along the lines of Walters *et al.*, 1993), and at one point in the workshop there was nothing for the fishermen to do while the biologist participants were off to do data analysis. A fisherman suggested that we let them play with an early version of the model, and I agreed provided they were careful not to take this version seriously. So they went away to play, and returned to my surprise within the hour with two things. First was a set of questions about why the model made some predictions that they found surprising (consequences of basic yield per recruit theory when harvest rate is reduced). Second was a whole set of very exciting ideas about how to use reduced fishing seasons, changes in timing of fishing, and spatial refuges to maintain their economic performance while substantially reducing exploitation rates (and hence increasing annual egg production and reducing risk of recruitment overfishing). I frankly found this outcome somewhat embarrassing, because the options that the fishermen identified on their own were ones that we (the biologists) should have been suggesting to them in the first place. If this accidental interaction is any indication of what can be obtained from more carefully planned and structured interactions between management players, the future for game playing approaches is indeed bright.

9.6 PITFALLS: MEGAMODELS, MEGAINFORMATION AND MEGAMISTAKES

There is a real danger that we will use new information gathering and modeling capabilities to grossly overparameterize models that are used for estimation of basic quantities such as optimum harvest rates and sustainable yields. Most biologists still assume that more realistic and detailed models are always preferable to simple ones when reasonable data are available to estimate parameters for the realistic ones. The basic fallacy of this argument was first demonstrated clearly (Hilborn, 1979; Ludwig and Walters, 1985) by generating fake population and harvest data from realistic models and then showing that these simulated populations could often be managed better by estimating management parameters (e.g. maximum sustainable yield (MSY) and optimum effort) for them using simpler, deliberately incorrect models. These simulations showed that the usefulness of a model for management analysis depends in a complicated way on how informative the data are for estimating its parameters, and good estimation performance can be more important than the biases introduced by deliberate simplifications in the model structure. Figure 9.2 shows a simple way to understand why it is often wise to deliberately choose the 'wrong' model. When a model is used to estimate a parameter of management relevance, such as MSY, there is a cost associated with error in this estimate when it is used by managers (missed harvest opportunity, or overharvest); this cost is zero only if the data happen to give exactly the right parameter estimates. For any model and data quality (types available, precision), there is a statistical distribution of possible error in the parameter estimates. For more realistic models, the estimator is usually unbiased (distribution centered on correct value), but can be quite inaccurate due to overparameterization (trying to estimate too many parameters from the data, so that each cannot be estimated accurately because its effects are confounded with effects of other parameters being estimated). Simpler models typically give more precise (less variable), but often biased estimates. The classic example of a very precise but biased model for MSY estimation is the Gulland or Fox equilibrium regression of cpue on effort; such equilibrium models give very precise estimates, but these are very often dangerously biased upward. An ideal model for estimation is one that has relatively low bias in a safe direction (e.g. underestimates MSY), but few enough parameters to allow estimation with reasonable accuracy. The best such balance cannot be predicted *a priori* just by looking at the data available or the biology of the organism involved; it must be discovered through reference model simulation methods that simulate data close to the actual case in point, and trial and error testing of alternative assessment models using this reference model (Hilborn and Walters, 1992). Much more such estimation testing work needs to be done in fisheries assessment, and this will continue to be a major area for fisheries computer applications.

Many of the important advances in computer data gathering and analysis that are opening to us now are mainly going to be valuable in measuring relative abundance and stock size trend more precisely, by accounting better for the spatial structure of abundance patterns. However, we have found in simulation studies that inaccuracies in policy parameter estimates from realistic models do not arise just because of inaccurate abundance data; indeed, simpler models can often be shown to perform better than realistic ones even when the simulated assessment system is given absolutely accurate (exact) abundance information. Rather, parameter inaccuracies are much more importantly affected by lack of informative contrasts in the historical data. For example, if stock size has not been reduced enough to provide direct empirical evidence of the reduction needed to impair average recruitment, any model for estimating that reduction is necessarily going to provide an imprecise estimate no matter how elegantly or realistically it may represent the details of the recruitment process. In short, it is not just more data of all sorts that we need for better fisheries modeling; we also need to make the best use possible of information from those unhappy circumstances where such parameters as the recruitment curve slope at low stock size have been discovered the hard way, through fishery collapse.

Perhaps the most fundamental issue and debate in fisheries today is about the relative importance of environmental factors versus factors related to stock size in causing fisheries collapses and policy failures. Major declines in groundfish recruitment off the east coast of North America appear to be related to declines in spawning stock, but could also be due at least in part to climatic changes in recent decades (Cushing, 1982; Mann, 1992; Myers *et al.*, 1993). Declines in survival rates of hatchery salmon populations have been attributed to environmental factors by supporters of hatchery programs, and to overstocking (natural carrying capacity effects) by detractors of the programs (Emlen *et al.*, 1990; Walters, 1993). These 'Thompson–Burkenroad debates' must be resolved somehow if we are to proceed wisely in programs to rebuild and maintain natural production systems (Walters and Collie, 1988). Will high technology environmental monitoring and computer simulation methods allow us to resolve the debates by finally understanding how environmental factors really have (and will) influenced population and survival declines? Better monitoring and modeling technologies will certainly allow us to construct more elaborate and detailed mechanistic models and hypotheses about how fish relate to their environment. But the pitfall here is that just because we can build more credible models is no guarantee at all that we will be able to critically and scientifically test the assumptions underlying these models. In particular, being able to relate survival and growth during some life stages, places, and times precisely to environmental factors does not in any way preclude the possibility that compensatory responses occur at other stages to override such relationships. That is, lovely mechanistic models can be dangerously incomplete, yet lull us into believing them through their very realism and good agreement with available data. In the end, no complex computer modeling or data gathering system can substitute for the fundamental scientific requirement of trying to challenge and reject our models through real field experience and experiments. Fisheries managers and funding agencies are not being told about this pitfall by the large community of oceanographers and fisheries scientists who have flocked to the environmental effects banner for funding. We may be able to mesmerize the managers with our elaborate models and data systems, and give them excuses not to take strong regulatory action when it is needed, but ultimately nature is not going to forgive us for this game.

A simple way to summarize the previous paragraph is to warn that we cannot substitute technology for sound scientific method. But one might counter this with the argument that when computerized measurement and analysis methods (along with careful biological process research) allow us to see how fish interact with their environment in greater detail, key patterns and relationships will just leap out at us and be so obvious as to not require critical testing. Indeed, this may well be the case with some relatively simple processes like movement, and how migration/distribution patterns are shaped by response to environmental factors. But it is dangerously misleading when applied to the most crucial and complex process in fisheries science, recruitment. There is a long tradition in recruitment studies of finding spurious correlations that break down the year after you publish them, and having more data and sophisticated analysis methods could just speed this tradition along. In fact, for most fish stocks we cannot do even the simplest possible test of elaborate recruitment–environment models, namely to see if they can reproduce historical recruitment data. For example, we recently completed an individual-based modeling exercise on groundfish stocks in the Hecate Strait BC. We began the analysis by using environmental databases and a fairly detailed hydrodynamic model to reconstruct (hindcast) historical current and thermal regimes in the Strait.

Then we introduced simulated larvae into these regimes for 1950–90 using a Walters *et al.* (1992) model. We found only very weak correlations between larval distribution/survival patterns predicted by this model and actual recruitment histories estimated from fisheries statistics. We tried to argue our way around this 'failure' by invoking the possibility of compensatory mortality at later juvenile stages, but a wise reviewer pointed out that we are not even entitled to this excuse. The reviewer pointed out that, as is the case for most fish populations, we simply did not have detailed enough historical data on spawning times and locations to even place larvae properly into our reconstructed physical fields. So our environmental pattern analysis and process modeling effort could not even get past the first life history stage leading to recruitment, in terms of the most basic validity check (past data) imaginable. And the really hard parts come later, in dealing with juvenile fish that have much more complex behaviors and reactions to environmental variables such as predation risk. We could, of course, just keep blundering along with the analysis while starting to gather the more detailed spawning data needed to check the larval stage calculations; but I strongly suspect that the end of this would be several lifetimes of finding new data needs, without ever a credible and well-tested result of value in fisheries recruitment forecasting or to help resolve Thompson–Burkenroad debates.

The pitfalls discussed in this section are not just ones for computer modeling freaks like myself to worry about. There is no reason to expect that the eyeball and armwave school of intuitive fisheries analysis is going to find more robust ways to analyze data so as to avoid fundamental errors of overparameterization, confusing availability of details with overall understanding, and basing policy on plausible hypotheses that fit the facts well but are dangerously misleading. Computers are offering all of us the chance to turn fisheries from a discipline of guesswork, dogma, and armwaving into one that really deserves to be called a science; we must all learn to use them wisely.

REFERENCES

Cushing, D.H. (1982) *Climate and fisheries*, Academic Press, London.

Deriso, R.B. (1985) Risk averse harvesting strategies, in *Resource Management, Proceedings of the Second Ralf Yorque Workshop*, (ed. M. Mangel), Lecture Notes in Biomathematics No. 61, Springer-Verlag, Berlin, pp. 65–73.

Deriso, R.B., Quinn, T.J. and Neal, P.R. (1985) Catch-age analysis with auxiliary information. *Canadian Journal of Fisheries and Aquatic Sciences*, **42**, 815–42.

Emlen, J.M., Reisenbichler, R.R., McGie, A.M. and Nickelson, T.E. (1990) Density-dependence at sea for coho salmon (*Oncorhynchus kisutch*). *Canadian Journal of Fisheries and Aquatic Sciences*, **47**, 1765–72.

Fournier, D.A. and Warburton, A.R. (1989) Evaluating fisheries management models by simulated adaptive control – introducing the composite model. *Canadian Journal of Fisheries and Aquatic Sciences*, **46**, 1002–12.

Hilborn, R. (1979) Comparison of fisheries control systems that utilize catch and effort data. *Journal of the Fisheries Research Board of Canada*, **36**, 1477–89.

Hilborn, R. and Walters, C. (1992) *Quantitative fisheries stock assessment and management: choice, dynamics, and uncertainty*. Chapman & Hall, New York.

Hutchings, J. and Myers, R.A. (1994) What can be learned from the collapse of a renewable resource? Atlantic cod, *Gadus morhua*, of Newfoundland and Labrador. *Canadian Journal of Fisheries and Aquatic Sciences*, **51**, in press.

Ludwig, D. and Walters, C. (1985) Are age-structured models appropriate for catch-effort data? *Canadian Journal of Fisheries and Aquatic Sciences*, **42**, 1066–72.

Mann, K.H. (1992) Physical oceanography, food chains, and fish stocks: a review. *ICES Journal of Marine Sciences*, **50**, 105–19.

McGuire, T.R. (1991) Science and the destruction of a shrimp fleet. *Marine Anthropological Studies*, **4**, 32–55.

Methot, R.D. (1990) Synthesis model: an adaptable framework for analysis of diverse stock assessment data. *International North Pacific Fishery Commission Bulletin*, **50**, 259–77.

Myers, R.A., Drinkwater, K.F., Barrowman, N.J. and Baird, J.W. (1993) Salinity and recruitment of Atlantic cod (*Gadus morhua*) in the Newfoundland region. *Canadian Journal of Fisheries and Aquatic Sciences*, **50**, 1599–609.

Parma, A. (1993) *Retrospective catch-at-age analysis of Pacific halibut: implications on assessment of harvesting policies*. Proceedings of the International Symposium on Management Strategies for Exploited Fish Populations, Alaska Sea Grant College Program, AK-SG-93-02, pp. 247–65.

Parma, A.M. and Deriso, R.B. (1990) Experimental harvesting of cyclic stocks in the face of alternative recruitment hypothesis. *Canadian Journal of Fisheries and Aquatic Sciences*, **47**, 595–610.

Pauly, D. (1994) *On the sex of fish and the gender of scientists*, Fish and Fisheries Series 14, Chapman & Hall, London, UK.

Pearse, P.H. and Walters, C. (1992) Harvesting regulation under quota management systems for ocean fisheries: decision making in the face of natural variability, weak information, risks and conflicting incentives. *Marine Policy*, **16**, 167–82.

Walters, C.J. (1993) *Where have all the coho gone?* pp. 1–9, in Proceedings of the coho workshop, Nanaimo BC, May 26–28, 1992, (eds L. Berg and P. Delaney), Dept. Fisheries and Oceans, Communications Directorate, Vancouver, BC.

Walters, C.J. (1994a) Lessons for fisheries assessment from the northern cod collapse. *Canadian Journal of Fisheries and Aquatic Sciences*, submitted.

Walters, C.J. (1994b) Using gaming procedures in the design of adaptive management policies. *Canadian Journal of Fisheries and Aquatic Sciences*, in press.

Walters, C.J. and Collie, J.S. (1988) Is research on environmental factors useful to fisheries management? *Canadian Journal of Fisheries and Aquatic Sciences*, **45**, 1848–54.

Walters, C.J., Hall, N., Brown, R. and Chubb, C. (1993) Spatial model for the population dynamics and exploitation of the Western Australian rock lobster, *Panulirus cygnus*. *Canadian Journal of Fisheries and Aquatic Sciences*, **50**, 1650–62.

Walters, C.J., Hannah, C.G. and Thomson, K. (1992) A microcomputer model for simulating effects of physical transport processes on fish larvae. *Fisheries Oceanography*, **1**, 11–19.

Walters, C.J. and Parma, A.M. (1995) Fixed exploitation rate policies for responding to climate impacts on recruitment. *Canadian Journal of Fisheries and Aquatic Sciences* (in press).

Walters, C.J. and Pearse, P. (1994) Uncertainty and options for insuring sustainability of quota management systems. *Reviews in Fish Biology and Fisheries*, in press.

Author index

Species index